CAMBRIDGE COUNTY GEOGRAPHIES

General Editor: F. H. H. GUILLEMARD, M.A., M.D.

T0352323

NOTTINGHAMSHIRE

Cambridge County Geographies

NOTTINGHAMSHIRE

by

H. H. SWINNERTON, D.Sc., F.Z.S., F.G.S.

With Maps, Diagrams and Illustrations

Cambridge:

at the University Press

1910

CAMBRIDGE UNIVERSITY PRESS
Cambridge, New York, Melbourne, Madrid, Cape Town,
Singapore, São Paulo, Delhi, Mexico City

Cambridge University Press
The Edinburgh Building, Cambridge CB2 8RU, UK

Published in the United States of America by Cambridge University Press, New York

www.cambridge.org
Information on this title: www.cambridge.org/9781107669789

First published 1910
First paperback edition 2013

A catalogue record for this publication is available from the British Library

ISBN 978-1-107-66978-9 Paperback

PREFACE

IN writing this book every effort has been made to secure accuracy; nevertheless, as errors may have crept in, corrections and suggestions will be gladly welcomed.

I am deeply grateful to the many who have placed their time and special local knowledge at my disposal, especially to Professors J. W. Carr and F. S. Granger, Rev. H. V. Dawes, Councillor E. Richards, J.P., Messrs A. T. Metcalfe, F.G.S., J. M. Stenton, M.A., F. Rayner, F. W. Davies, G. Goodall, B. Smith, M.A., and R. F. Percy.

H. H. SWINNERTON.

UNIVERSITY COLLEGE,
  NOTTINGHAM.
    July, 1910.

CONTENTS

ILLUSTRATIONS

1. Nottingham and Nottinghamshire. Meaning and Origin of the Words. Shire and County.

A visitor is not long in Nottingham before he hears of the caves in the Castle Rock and elsewhere, and is taken to see them. To-day they are merely curiosities, but in bygone years some of them were used as dwellings. The earliest historic notice of Nottingham refers to the place under the British name Tigguocobauc, which means "the house of caves." From this has arisen the erroneous impression that the present name also has the same meaning.

Before it was modified by Norman influence the name had several forms, e.g. Snothryngham, Snottingaham, Snottingham—but Snotengaham was the earliest. This ending "ham" is akin to the word home, and is of Anglo-Saxon origin. It tells us of a people who came to this country—not, as the Romans did, to exploit—but to colonise and to make for themselves a home. It is not at all unlikely that Snottingham was the home of an Anglian family—Snot (the wise) by name. Thus with the possessive " ing " the whole word means " The home of Snot."

Caves in Rock Cemetery, Nottingham

With the City of Nottingham is associated a large extent of country known by the interchangeable names of Nottinghamshire and the County of Nottingham. These also reflect the influence of the Anglo-Saxons and Normans respectively. The suffix "shire" is akin to the word "share," and like it signifies a division, something cut off. "County" only dates from Norman times and denotes the domain of a *Comte* or Count. There is no historical account of the coming into existence of Nottinghamshire. At one time the area it covers seems to have been divided between the kingdoms of Mercia and Lindsey. It certainly existed as a shire before 1016, for the word Snotingahamscire occurs in writings of that date. Probably it was created a century before by Edward the Elder. This king in order to consolidate the newly acquired portion of his kingdom, situated in the region now called the Midlands, placed "shares" of it under the control of chief men or Ealdormen. Usually each share or shire, as it was called, consisted of those portions of the country that were within easy reach of a military centre. Nottingham was such a centre, situated not far from the old Roman Fosseway, and at a point where an important road to the north crossed the Trent. It thus dominated the country through which these three ways passed. That country was naturally made into a shire administered from Nottingham and therefore called Nottinghamshire. On the other hand the shire may have been formed from the district which was settled by that Danish army which had its head-quarters in Nottingham.

As has just been remarked, the expressions County of

Nottingham and Nottinghamshire are used indifferently. But shire and county do not quite mean the same thing. Though we speak of all the divisions of England collectively as counties not all of them are strictly so—as for example the Duchy of Cornwall—nor is the suffix shire applied except to such divisions of our land as once formed part of the early English kingdoms. There are other counties, such as Kent, Essex, Suffolk, etc., which reveal their origin in their names—the kingdom of the Cantii, of the East Saxons, of the people south of the Little Ouse and Waveney—and these, it should be observed, have in most cases kept their boundaries unaltered to this day.

As a rule, however, our counties are by no means fixed quantities, and their boundaries—even in spite of the fact that they are mainly dependent on the natural delimitation of river or range, and not as in America marked out with a ruler—have often altered. They are, in fact, altering now, and the county of London has come into existence within the recollection of all of us.

2. General Characteristics.

Nottinghamshire has no seaboard. Nevertheless those reaches of the Trent which form part of its eastern boundary are influenced by the tide and can carry ships of 150 tons burden. Throughout the length of the county the Trent has always been an important highway and a dominant factor in its history and commercial development. During the nineteenth century the river became overshadowed by the railways and its traffic declined. Recent years have brought with them a renewed interest in it which is leading to a realisation of its possibilities as a commercial highway.

Thus in spite of its inland position this county has some of the advantages of a seaboard. In other respects it exhibits a mixed character. To the north-west lies

Nottingham from The Meadows

upland England, with regions rich in mineral wealth and thickly peopled with an industrial population. To the south-east is lowland England, pastoral and agricultural, and dotted with residences and parks. Nottinghamshire lies on the boundary between the two and participates in the characteristics of both.

Its highest ground is on the west and is the extremity of a spur from the Pennines. Throughout the whole of its eastern border the ground is very low, never reaching an altitude of 100 feet.

The higher ground of the west furnished a few streams capable of supplying water-power for driving machinery during the early days of the industrial revolution. In the same part of the county the presence of abundance of coal served still to retain the industries when water-power was replaced by steam.

In the east no such natural advantages existed to favour the development of industries. That part of the county therefore remains purely pastoral and agricultural. In the heart of the county palatial residences and extensive parks are so numerous that this region is often called " The Dukeries."

3. Size. Shape. Boundaries.

Nottinghamshire is one of the North Midland counties. It is bordered on the north-west by Yorkshire, on the west by Derbyshire, on the south by Leicestershire, and on the east and north-east by Lincolnshire. The most northerly point is situated at the junction of its boundaries with Yorkshire and Lincolnshire. The most southerly point is where the Fosseway enters the county. The line joining these two is the longest in the county, and extends 51 miles. The greatest width is 27 miles, along a line running close to Selston and Newark.

The area of the ancient or geographical county is 539,756 acres or 843 square miles. These figures include

3426 acres of water. In area this county occupies a central position among the English and Welsh counties, of which 25 are larger, and the same number smaller. It is one-seventh the size of Yorkshire, and six times larger than Rutland. It is practically equal to Herefordshire, which has an area of 842 square miles.

Creswell Crags

(*Derbyshire to left, Notts to right of gap*)

The general shape of the county is that of an oval, with its long axis lying nearly north and south. The circumference of this oval is about 120 miles, but numerous irregularities increase the length of the actual boundary to over 180 miles. One half of this is artificial, the other half is naturally defined by rivers and streams.

There are three long stretches of natural boundary. One starts close to the source of the Erewash in the extreme west, near Pinxton, and follows down this stream almost to the Trent. Another starts near to the mouth of the Erewash and follows up the Trent, the Soar, and one of its tributaries to near Rempstone. The third, in the north-east, follows down the Trent from North Clifton to West Stockwith. This one is broken by a slight deviation close to Dunham bridge, where one Nottinghamshire field lies on the Lincolnshire side of the river. This field is the nearest approach to a detached portion of which the county can boast. Close to Bole the Trent formerly flowed round two remarkable loops with narrow mouths opening into Lincolnshire. The land thus nearly surrounded by water belonged naturally to that county. In 1797 the river assumed its present course across the narrow necks of the loops. It is only on the latest maps that this land is represented as belonging to Nottinghamshire.

4. Surface and General Features.

Our county lies in the southern portion of the large plain whose drainage falls into the Humber. This portion is called the Vale of Nottingham. It is bounded on the west by the Pennines, and on the east and south by the Lincoln Cliff and its natural continuation the Belvoir Escarpment. These boundaries all lie outside the county but nevertheless have greatly influenced its history and commerce.   At the south-west corner there is a gap of low country towards which the valleys of the Erewash, Derwent, Upper Trent, and Soar converge.   This gap is the natural gateway from western and southern England

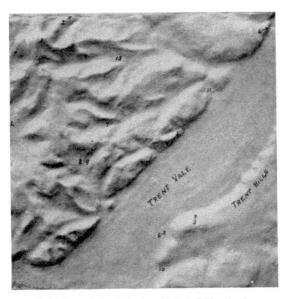

Relief-Map of the District N.E. of Nottingham

1. Cockpit Hill.   6. View point, p. 14.   7. Lambley.
8. Gedling Colliery.   10. Radcliffe.   11. Lowdham.
13. Woodborough

into Nottinghamshire and thence to the north.   Three minor gaps leading from Lincolnshire and the south occur at Lincoln, Ancaster, and Grantham.

The county is largely lowland in character. Only one fifty-sixth of its area is above the 600-feet contour, and no less than two-fifths is below 100 feet in altitude. Generally, as we have seen, the lower ground is in the north and the east; the higher in the south-west. The latter is formed partly by a continuous range of hills which flanks the Pennines, but is separated from them by the valleys of the Erewash and Rother. These hills slope steeply westwards and gently eastwards. The highest ground in the county is situated in the district round the Robin Hood's Hills, just where this range is joined by the watershed between those two rivers. It culminates near Hucknall-under-Huthwaite in a height of 651 feet. North of this the range lies outside the county, southward it passes as a gradually declining spur which disappears west of Nottingham. This range will be afterwards referred to as the Western Hills.

The highest ground south of the Trent is formed by a northerly spur of the Belvoir Escarpment. This is known as the Wolds and dominates this corner of the county.

A broken range of hills commences in the eastern suburbs of Nottingham, and sweeps northwards along the central axis of the county, ending abruptly at Gringley-on-the-Hill. Its course thus lies parallel to and 15 miles from the Belvoir Escarpment and the Lincoln Cliff. The detached portions of these central hills decrease in height from 580 feet at Cockpit Hill to 253 feet at Gringley. From their summits the ground falls rapidly to the west and very gently towards the east. The

Vale of Belvoir from Green Hill near Wartnaby

(Plain formed by Lower Lias Clays and Limestones)

western outlook from Cockpit Hill is occupied mainly by the Western Hills, that from Gringley by an extensive area of low country, the Vale of the Idle. The eastern outlook is, everywhere, over a plateau which slopes gradually until it is lost at the fringe of the county in the Vale of Belvoir and the plain of the Trent.

Between Nottingham and Newark the continuity of the plateau is broken by the Trent Vale, which here cuts across it like a great trench two miles wide (see p. 9). The hills which bound the vale on the south-east are called the Trent Hills. The surface of the plateau is everywhere beautifully diversified with little "dumbles" and valleys which open on to the vale. The whole presents a fair example of a miniature dissected plateau.

North of Gringley lie the flat Carr lands or fenlands. These, together with portions of the valleys of the Idle and the Trent, were at one time swamps. Though this country is now well drained and cultivated, the memory of its former condition is still perpetuated in such names as Mattersey (*ey* = an island), and Misterton Soss (*Soss* = a dirty puddle).

Up to the close of the sixteenth century the Royal Forest of Sherwood had an area of about 100,000 acres. At that time its borders extended from Nottingham up the Leen to Newstead, thence through Kirkby and Skegby to the Meden and along that stream to Perle-thorpe; from this place it stretched southwards along the North Road, Dover Beck, and the Trent back to Nottingham. It consisted of open wastes and woodland and was a favourite hunting-ground of the early kings.

In later times the timber from its oaks was in great demand and was used, among other purposes, for the Navy and in the construction of St Paul's Cathedral. Many of the great parks in the county are enclosed portions of this ancient forest.

5.   Rivers—(*i*) The Trent.

With the exception of two small portions, Nottinghamshire lies wholly within the basin of the Trent.   The basin of the Rother touches it north of Sutton-in-Ashfield, and the basin of the Witham just east of Newark.   The Trent is 147 miles long, and is second in length only to the Severn amongst English rivers.   When it enters the county it has already received the drainage from two-thirds of its basin, which has an area of 4052 square miles.   It is then a fine river with clear water and a rapid current. Its general course through the county is shaped like a bow with its convexity towards the east.   Throughout it lies in a narrow vale about two miles wide.

For the first third of its course the vale is trench-like and clearly defined by hills on either side.   On the west these rise to over 300 feet; but they gradually decline until at last the vale proper is almost indistinguishable from the low country through which it passes for the remainder of its course.

The windings or meanders of the river are of two types.   The larger ones sweep from one side of the vale to the other (see p. 15) and are sometimes three miles from bend to bend. These are probably inherited from the times when the Trent was a fuller stream than it is now. Superposed on these are smaller meanders which are usually less than one mile wide. These belong to the Trent as we now know it.

Though the river swings from one side of the vale to the other it "hangs" to the right. This is shown by the

Malkin Hill and the Trent below Radcliffe

fact that it is only on this side that the river impinges against the base of the hills which bound its vale, as shown in the illustration above. Here it has undercut them, and their usually gentle slopes are replaced by precipitous cliffs. At these places the river is seen in the act of widening its vale. It is therefore interesting to note that those villages whose names (e.g. Radcliffe) are based upon

cliff as the predominant natural feature of the neighbourhood are all confined to the right side of the vale.

The height above sea level of the vale at its entry into
the county is only 90 feet; where it leaves it it is about
15 feet. The land level is usually only a few feet above
the river-water level. When, therefore, the rainfall is
heavy the river is unable to carry off immediately the

View of the Vale of Trent from Malkin Hill

excess of water, and the surrounding country becomes
flooded. In former years these floods were often very
serious. The greatest of which we have record occurred
in 1875. They are more frequent in the lower than
the upper reaches, because there the river flows more
slowly and is also subject to tidal influence.

In the lower reaches the flow of water is stopped and

completely reversed twice every twenty-four hours by the tide. The influence of the spring tide is felt as far as Sutton, but for some miles above Stockwith it is shown as a remarkable "bore." This is best seen at spring tide, when the level of the water is raised as much as five or six feet in a short time. The bore rushes up the river

The Aegir advancing up the Trent near Gainsborough

as a wall of water followed by a series of waves, and brings sudden disaster to any boat it finds unprepared. The wall is called "the Aegir," the waves are the "whelps." This former name survives "like the peak of a submerged world" and tells us of those old Norsemen who recognised in the Aegir their god the "Sea-Tempest."

6. Rivers—(*ii*) The Tributaries of the Trent.

Within the county the Trent receives many more tributaries on its left than on its right bank.

A glance at the map shows that those on the left may be classified into long streams which rise on the

The Meden at Budby

(*Typical Bunter Sandstone-stream*)

Western Hills, and short ones which rise on the Central Hills. To the former belong the Erewash, the Leen, the Dover Beck, and the Idle with its tributaries the Rainworth Water, Maun, Meden, Poulter, and Ryton. The last two rise outside the county; the remainder on either side of a short watershed which trends from west to east through the Robin Hood's Hills. From this a minor water-parting runs southwards and separates the basins of the Erewash and the Leen. The latter has on its left side a solitary tributary, the Daybrook, which rises on the Central Hills.

A long strip of the county is characterised by the fact that it gives rise to very few streams. It is narrowest at Nottingham. Thence it widens northwards until it occupies nearly the whole of the north-west quarter of the county. The streams which cross it exhibit a general absence of tributaries.

The longest watershed is one which passes along the crests of the Central Hills. A few small streams flow westwards from this line and fall into the Leen, Maun, and Idle. Many more flow eastward. In the upper parts of their courses they are little torrents and have made for themselves waterfalls and gorges. The latter are sometimes 20 to 30 feet deep and are known as "dumbles." They lie at the bottoms of broad open valleys and are hidden from view by overhanging trees.

To the right of the Trent the main water-parting runs north and south along the Wolds. From Newark to the point where this meets the Trent there are no tributaries. All the drainage from the area east of the divide is collected by the Smite and the lower reaches of the Devon. Both these streams rise in Leicestershire, but the former runs for the greater part of its length through Nottinghamshire. From the other side of the Wolds the streams at first flow westward. One continues in this

The River-System of Nottinghamshire

(*Showing every tributary-stream and dyke, and the marked contrast between the Sandstone country on the west and the Clay country on the east*)

direction and falls into the Soar, the remainder turn sharply to the north and enter the Trent.

There are no natural lakes in the county. One known as the Whitewater formerly existed just south of Styrrup, but it was drained and now forms Whitewater Common. Several artificial lakes exist on the courses of the Maun, Poulter, and Rainworth Water.

7. Geology and Soil.

By Geology we mean the study of the rocks, and we must at the outset explain that the term *rock* is used by the geologist without any reference to the hardness or compactness of the material to which the name is applied; thus he speaks of loose sand as a rock equally with a hard substance like granite.

If all the soil were stripped from the surface of the county there would be found underlying it several kinds of rocks. These occupy the areas shown by the colours on the geological map. To study these rocks, however, it is only necessary to examine them in pits, quarries, and railway cuttings. Their extent can be discovered by observing changes in the nature of the soil and the shape of the ground.

The Permian area is occupied mainly by Magnesian Limestone, so-called because it contains magnesium carbonate. Along its eastern border a narrow strip of deep red clay, Permian marl, exists but it is not separately indicated on the map.

	Names of Systems	Subdivisions	Characters of Rocks
TERTIARY	Recent Pleistocene	Metal Age Deposits Neolithic ,, Palaeolithic ,, Glacial ,,	Superficial Deposits
	Pliocene	Cromer Series Weybourne Crag Chillesford and Norwich Crags Red and Walton Crags Coralline Crag	Sands chiefly
	Miocene	Absent from Britain	
	Eocene	Fluviomarine Beds of Hampshire Bagshot Beds London Clay Oldhaven Beds, Woolwich and Reading Thanet Sands [Groups	Clays and Sands chiefly
SECONDARY	Cretaceous	Chalk Upper Greensand and Gault Lower Greensand Weald Clay Hastings Sands	Chalk at top Sandstones, Mud and Clays below
	Jurassic	Purbeck Beds Portland Beds Kimmeridge Clay Corallian Beds Oxford Clay and Kellaways Rock Cornbrash Forest Marble Great Oolite with Stonesfield Slate Inferior Oolite Lias—Upper, Middle, and Lower	Shales, Sandstones and Oolitic Limestones
	Triassic	Rhaetic Keuper Marls Keuper Sandstone Upper Bunter Sandstone Bunter Pebble Beds Lower Bunter Sandstone	Red Sandstones and Marls, Gypsum and Salt
PRIMARY	Permian	Magnesian Limestone and Sandstone Marl Slate Lower Permian Sandstone	Red Sandstones and Magnesian Limestone
	Carboniferous	Coal Measures Millstone Grit Mountain Limestone Basal Carboniferous Rocks	Sandstones, Shales and Coals at top Sandstones in middle Limestone and Shales below
	Devonian	Upper } Mid } Devonian and Old Red Sand- Lower } stone	Red Sandstones, Shales, Slates and Lime- stones
	Silurian	Ludlow Beds Wenlock Beds Llandovery Beds	Sandstones, Shales and Thin Limestones
	Ordovician	Caradoc Beds Llandeilo Beds Arenig Beds	Shales, Slates, Sandstones and Thin Limestones
	Cambrian	Tremadoc Slates Lingula Flags Menevian Beds Harlech Grits and Llanberis Slates	Slates and Sandstones
	Pre-Cambrian	No definite classification yet made	Sandstones, Slates and Volcanic Rocks

In the next area the rock is a soft sandstone. It is often bright red, variegated with greenish bands; hence the name Bunter (Germ. *bunt* = variegated). In many places it contains an abundance of pebbles, usually of quartzite. It is a very porous rock, and rain-water therefore soaks into it easily instead of running off. This is the reason why so few streams have their source in this part of the county. An exception occurs in the case of the tributaries of the Erewash and Leen. The hills in which they rise are capped with this sandstone resting on a bed of Permian marl. The water cannot sink from the stone into the marl and so it runs out at the margin and forms the sources of these streams.

In the Waterstones area there are alternating bands of sandstone and clay.

The Keuper marl presents a marked contrast to the Bunter. It consists of a bright red clay with here and there an insignificant layer of sandstone called "skerry." When rain falls, a little soaks into the skerry but the greater portion flows off at once and forms the streams which are so characteristic of this part of the county. Near its eastern margin and between Laxton and Askham this clay contains layers of gypsum.

In the Lias area, blue-grey limestone is found bordering the Keuper; the remainder is occupied by clay.

In 1876 a boring was made near Collingham in the extreme east of the county. It penetrated the ground to a depth of over 2000 feet. Starting in the Lias it passed successively through rocks identical with the Keuper marl, Waterstones, Bunter, and Permian. This shows that the rocks seen at the surface slope one under another towards the east. In fact they form extensive layers lying one on the top of another. The area over which one of these layers comes to the surface is called the outcrop. The coloured areas on the map show the position and extent of the outcrop of each layer.

Waterstones: Sherwood

(*Freshly cut surface showing alternating bands of sandstone and clay*)

The Permian layer, unlike those above it, is thickest in the north. As it passes southwards, it gradually becomes thinner until it eventually disappears entirely near Radford.

These layers of course extend beyond into the surrounding counties, where many other layers occur. The presence in these rocks of such things as fossils and ripplemarks shows that nearly all have been laid down in water. Now suppose that a trench or cutting, about 1000 feet deep, were made across England, the edges of these layers would appear in the sides of this trench, and the layers, or strata as they are called, would be seen to overlap one another like pennies in a pile which has fallen down. If the coins were carefully slipped back into place once more the pile thus formed would illustrate the way in which the table of systems here given has been compiled. If such a pile were made out of the rocks themselves it would be between 10,000 and 20,000 feet high. The rocks represented at the bottom of the table are the oldest, those at the top are the youngest. For convenience of reference they are divided into three main groups; Primary or Palaeozoic, Secondary or Mesozoic, Tertiary or Cainozoic. They rest upon a foundation of still older rocks which may be spoken of as the Precambrian. These main divisions are further subdivided into systems.

The names of these systems are arranged in order in the table. On the right hand side, the general characters of the rocks of each system are stated.

When these rocks were first laid down in water they were flat and horizontal. Afterwards some of them became tilted, as for example the rocks already described above. In other counties, as for instance Derbyshire, the layers have been folded into arches and troughs. When this folding started in Derbyshire the Permian, Bunter, and later rocks had not been formed, and the Carboniferous Limestone, which may be seen to-day in the beautiful dales of that county, was buried under a thick covering of grey clays, dark shales, and light-coloured sandstones with occasional coal seams. This covering, which is now spoken of as the coal-measures, then extended continuously from Lincolnshire to North Wales. When the folding began the coal-measures were swept off the tops of the arches by the action of the rain, frost, rivers, and streams. Thus the underlying limestone was exposed and raised to form the Derbyshire Pennines. Meanwhile the coal-measures were left in the saucer-like troughs on either side and were afterwards buried under the Permian and later rocks. The eastern saucer is still largely buried out of sight, but its western margin outcrops in Yorkshire, Derbyshire, and West Nottinghamshire and assumes a south-easterly trend at Wollaton, where it disappears under the Bunter. This outcrop is the visible coalfield. The buried portion is the concealed coalfield.

In other counties, as for example in our neighbour Leicestershire, another kind of rock is found which has not been formed in water. At one time it was situated deep down in the earth and was so hot that it was molten. In this condition it was forced towards the surface. Sometimes it cooled before it reached the surface, at other times it found an outlet and then it produced great volcanoes and was poured forth as lava. Rocks formed in these ways are called igneous rocks. Though they do not occur naturally in Nottinghamshire they may often be seen used as setts, curbs, and pillars.

All these rocks together are the block out of which rain, frost, and streams have sculptured the surface features of the county. The Magnesian Limestone has resisted the carving action of these agents. Coming up from beneath

Keuper Clay in Mapperley brickyards

the Bunter it rises steadily towards the west. With a capping of sandstone here and there it forms the water-shed of the Western Hills. The Keuper marls have not resisted to the same extent. The surface has been planed flat and has formed the gently-sloping plateau which extends from the Central Hills to the plain of the Trent and Vale of Belvoir. The western faces of these hills are formed by the edge of the more resistant Waterstones.

The Bunter produces much of both the highest and the lowest ground in the county. Where streams exist they have carved it easily. Where they do not exist, the rain though assisted by frost has had little chance, for it has been absorbed.

We may now turn to the soil which these varying strata afford, for it is the débris formed by the action of air, rain, frost, and plants upon these latter that composes it. For this reason the soil varies with the underlying rock. Where this is clay, the soil is clayey. Where it is sandstone with pebbles the soil is sandy and stony, or even gravelly. Where it is sandstone alternating with clay, a sandy or clayey loam exists. The limestone gives rise to a calcareous loam.

Most of the soils in the county lie where they were formed and therefore have the same character as the underlying rock. Some, however, differ greatly from this. These have been washed by the rain from soils of widely-separated districts, carried as sand, silt, or mud by the streams, and deposited to form alluvium by the rivers in time of flood.

The largest area of alluvium in our county is in the Trent Vale. Here it usually lies upon a coarse gravel, which was deposited before the river had dwindled to its present size. In some places the gravel is still not covered even by the highest floods. These gravel patches stand up like islands in the alluvial plain, as is shown in the illustration on p. 145.

In bygone ages glaciers and ice-sheets also transported soils, but only slight traces of their action are found in Nottinghamshire.

In places which, until modern times, were swamps and marshes the soil usually contains so much vegetable material as to be black and peaty. Occasionally shells are so abundant that a shell marl has been produced, as for example at Gotham.

8. Natural History.

The animals and plants of Great Britain closely resemble those of the Continent but they are not so numerous in species. For example, our islands possess only half the number of species of mammals found in Germany. To understand this it is necessary to go back in thought to a time when Great Britain formed part of a broad peninsula projecting from the north-west of Europe, and when the North Sea and English Channel were monotonous plains. At that time the climate was warm, and this country was just as rich in plants and animals as the rest of the Continent. Gradually the climate became arctic in severity, and the whole of the north-west of Europe became like Greenland, covered with snowfields and glaciers. All forms of life were forced to seek a haven in the more genial south. During this period the peninsula became partially submerged in the sea and consequently many of those plants and animals which still lingered immediately south of Ireland and England were also driven away.

Eventually the climate improved, the snowfields and glaciers retired to the uplands, retaining those forms of life that haunted their fringes. Thus it came about that many types peculiar to our mountain regions found their present home. Meanwhile the animals and plants in the south slowly spread back again over the north-west of Europe and began to find their way across the narrow connection then left between our islands and the Continent. But long before they had all returned so far, even this connection was lost. Since then it has never been restored for any great length of time, and consequently the British Isles have never been completely restocked.

It must not be imagined that immigration ceased with these conditions. The winds and birds still bring seeds and drop them here. Some of these seeds find a suitable habitat and become established. But the main factor in the introduction of species is man, who has been constantly bringing in fresh species. To him this country owes the importation of domestic animals and the food plants—wheat, potatoes, etc., and a number of stray plants accidentally introduced with commercial produce from other lands.

In Nottinghamshire there are 854 recorded species of wild plants. This is only half the number recorded for the whole of Great Britain. This small proportion is due largely to the inland position of the county and to the absence of uplands. In the adjoining counties there are extensive areas of limestone, the soil of which is peculiarly favourable to the growth of plants. In this county the richest flora, which includes such forms as the columbine, rock rose (*Helianthemum*), and bladder campion (*Silene cucubalus*), is found on the Magnesian Limestone, but this occupies only eight per cent. of the county. This area is also the richest in land snails, which of course require calcareous material for their shells.

Lime-loving plants, however, are not confined to the outcrop of the limestone, for some lime exists in the Keuper marls and Lias. Hence such plants as wood anemone, giant bell-flower, herb paris (*Paris quadrifolia*), and autumnal gentian (*Gentiana amarella*), are widely diffused through the county. Side by side with these occur clay-loving forms like the butterfly orchid (*Habenaria chloroleuca*), meadow geranium (*G. pratense*), and primrose.

The outcrop of the Bunter, which is three times as extensive as that of the Magnesian Limestone, is characterised by the poorness of its flora. The soil is so coarse that it does not draw up the water from below, nor does its surface hold much of that which falls upon it. The plants it bears, therefore, are such as prefer a dry and sandy situation, such as gorse, ling, heath, broom and hare's-foot (*Trifolium arvense*). Lime-loving plants do not thrive upon it.

It is on the Bunter that open commons and wastes, such as Bulwell Forest, occur, and furnish a suitable home for the common lizard, slow-worm, and viper.

On it also is Sherwood Forest, with its oaks and birches and monotonous undergrowth of bracken. Formerly the forest was much more extensive than it

is now and so was its fauna. Before the Norman in-
vasion it was the common hunting-ground of all. After
that it became the favourite preserve of kings and ecclesiastics. In those days the red deer, wolf, and badger were
common. Now the forest has become restricted and
enclosed. The wolf, wild boar, and wild cat have
become extinct. Red deer persist in a domesticated

Typical Bunter Country; bracken, gorse, and woodland

condition only on the great estates. The badger still
roams at night. Even now, however, many kinds of
birds and insects find a suitable home here. The modern
hunter is still rewarded from time to time by finding
some obscure insect, spider, or similar creature which has
never been recorded before for the county or possibly even
for Britain.

Birches in Birkland, Sherwood Forest: bracken undergrowth

Another asset to the Natural History of the county is the Trent. Its flood plain provides a home for sand-loving, clay-loving and moisture-loving plants; whilst its waters are rich in many kinds of snails, bivalves, and fishes. Sturgeon have been caught close to Nottingham. Porpoises and seals are occasional visitors and flounders are common in its tidal reaches. Lampern are still abundant.

In the immediate neighbourhood of Nottingham the Nottingham catchfly (*Silene nutans*) existed until quite recently in great abundance on the Castle walls and rock. The Meadows also were carpeted with crocuses within the recollection of the older inhabitants, but now their place is almost entirely taken by houses and their memory preserved in the names of streets. The draining of marshy land accounts for the scarcity or extinction of the sundew (*Drosera*), butterwort (*Pinguicula vulgaris*), and buckbean (*Menyanthes trifoliata*). On the other hand the importation of cereals and linseed from abroad has led to the introduction of other plants, e.g. bur parsley (*Caucalis daucoides*), and corn speedwell (*Veronica Tournefortii*), and various clovers and grasses, some of which have become vexatious weeds.

9. Reclaimed Land.

Many of our counties bordering on the sea have suffered, and are still suffering, very severe losses from the erosive action of the waves. In Norfolk and Suffolk this is very strongly marked and has become a very serious problem. Nottinghamshire, owing to its inland position, is not confronted as yet with any difficulties of this kind and no land has become lost to the county as the result of natural processes. On the other hand much has been reclaimed by man from a condition of swamp and marsh.

In the early periods of the county's history much of the land, especially in the north, was in such a condition. Wherever a river ran through low land the country was flooded from time to time. This was the case throughout the vales of the Trent, the Idle, the Smite, and the Devon. After the flood much water still remained on the land, converting it into perpetual marsh.

The country north of Gringley and along the lower reaches of the Trent was exposed to an additional cause of flooding. A large portion of it was only a few feet above the sea level. When the spring tides came in they raised the waters of the Trent and Idle as much as five or six feet. The land was therefore regularly covered with water at least once a fortnight. In some places this occurred twice every twenty-four hours.

The first efforts to reclaim this land seem to have been made by the Romans, for it was they who constructed the Bycar Dyke and the Fosse Dyke.

After this no serious attempts were made, except perhaps by the monks, until the reigns of Elizabeth and James I. During the reign of the latter, Dutch methods of draining were adopted in many parts of England, and great schemes were devised and carried through by the Dutch engineer Vermuijden. To his skill this county is said to owe the Morther Drain. But it is possible that this is to be identified with the Bycar Dyke of the Romans. It is deep and canal-like and runs alongside the Idle, but at a much lower level. The water from fourteen and a half square miles of fenland south of the Idle flows into minor drains and from them into the Morther Drain. About a mile up the Idle is Misterton

Tide-gates on the Idle at Misterton Soss

Soss. Here there is a pumping station which raises the water from the low level of the drain to a height sufficient to enable it to flow away into the Trent. North of the Idle the drainage system is connected with that of the extensive area in South Yorkshire and North Lincolnshire known as the Levels of Hatfield Chase.

Alongside the four rivers mentioned above, strong flood banks have been constructed to hold back the high waters

3—2

from the land. The tributary streams enter the rivers through tunnels in these banks. At the mouth of each tunnel is a strong door which opens only outwards. The streams are thus able to discharge their waters into the river; but the flood water merely closes the door, and is thus prevented from flowing through on to the land. These doors are called flood or tide-gates. The best example is seen at Misterton Soss (p. 35). Here great tide-gates stand in the arches of a bridge across the Idle. When the tide sweeps up the river it automatically closes the gates and the country beyond is saved from flooding.

The tidal water brings with it much silt from the Humber. Wherever the water rests for a while this silt is deposited, often to a depth of an inch during one tide. This "warp," as it is called, makes fertile soil. Occasionally therefore a farmer will allow the water to pass through the gates on to his land.

Advantage is also taken of this to make the tide mend damaged portions of the river's bank. Bundles of brush-wood are fastened in the place. The tide deposits silt between and on the brushwood and thus builds up a good bank as seen in the foreground on the right in the illustration on p. 16.

Finally, careful deepening of the channel of the Trent by dredging has made it possible for flood waters to flow away more rapidly. Many square miles of the most fertile land in the county have by these means been rendered available for cultivation.

10. Climate and Rainfall.

The average weather of any place is called its climate. This depends primarily upon its latitude. But places upon the same latitude may have different climates owing chiefly to the proximity of the sea. Other less important factors are the direction and strength of the winds, the amount of sunshine, the temperature, and the rainfall.

Most of our weather comes to us from the Atlantic. It would be impossible here within the limits of a short chapter to discuss fully the causes which affect or control weather changes. It must suffice to say that the conditions are in the main either cyclonic or anticyclonic, which terms may be best explained, perhaps, by comparing the air currents to a stream of water. In a stream a chain of eddies may often be seen fringing the more steadily-moving central water. Regarding the general north-easterly moving air from the Atlantic as such a stream, a chain of eddies may be developed in a belt parallel with its general direction. This belt of eddies, or cyclones as they are termed, tends to shift its position, sometimes passing over our islands, sometimes to the north or south of them, and it is to this shifting that most of our weather changes are due. Cyclonic conditions are associated with a greater or less amount of atmospheric disturbance; anticyclonic with calms.

The Royal Meteorological Society receives reports of the temperature of the air, the hours of sunshine, the rain-fall, and the direction and force of the wind from all parts

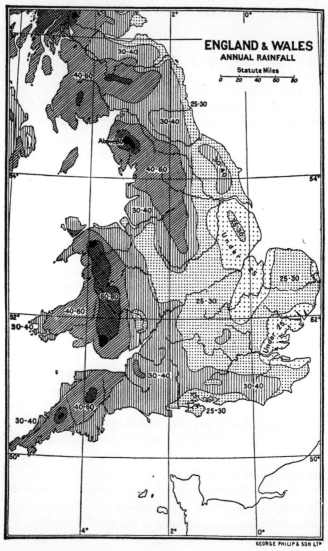

ENGLAND & WALES
ANNUAL RAINFALL

Statute Miles
0   20   40   60   80

30-40
40-60
25-30
30-40
Aberdeen
40-60
30-40
30-40
60-80
40-60
30-40
30-40
Under 25
25-30
25-30
Under 25
30-40
30-40
40-60
25-30

GEORGE PHILIP & SON LTD

(The figures give the approximate annual rainfall in inches.)

of our islands, and publishes a daily report with a map, which may be seen in almost any newspaper. We can at a glance learn what has happened in other parts of England during the last twenty-four hours. At the end of the year these results are all brought together and averaged, and we can compare this average climate of the different British areas.

In Nottinghamshire there are several stations, well equipped, where records have been kept for many years.

Of all the observations made upon the wind at Hodsock Priory during the thirty years 1876–1905, one-third showed the wind coming from the south-west and west; nearly as many indicated calms; the rest indicated winds coming from other quarters. The prevailing winds of our county, therefore, as of England generally, are from the south-west or west. The latter are most frequent in July, the former in December. Easterly winds are commonest in March.

In 1908, only 1228 hours of bright sunshine were recorded at Hodsock. This was less than the average for twenty-five years, viz. 1252. In the same year the sunniest month was May, with 170 hours of sunshine. The dullest was December, with only 30 hours. Thus the range was from 35 per cent. to 13 per cent. of the time the sun was above the horizon. These figures are approximately true for the rest of the county, which thus compares unfavourably with the south coast of England and some parts of the eastern counties seaboard. On the other hand it is better off than some of the uplands of Derbyshire and Yorkshire, which have an average of less than 1200 hours sunshine in the year, or than many great cities such as Manchester, where in 1907 only 894 hours were recorded.

On May 20th, 1909, the highest temperature reached in Nottingham was 66° Fahr.; the lowest was 40°. These observations exhibit a range of no less than 26° Fahr. within twenty-four hours. In the Scilly Isles on the same day the range was only 9°, i.e. from a maximum of 59° to a minimum of 50°. This difference between the two places is due to the fact that Nottingham is surrounded on all sides by land, the Scilly Isles by water.

It is well known that if the sun shines with equal strength on land and water, the former becomes heated more rapidly and to a higher temperature than the latter, while, after the sun has set, if the sky is clear, the heat radiates away from the land more rapidly than from the water. These two influences, sunshine and radiation, are most effective in the summer time. In July at Nottingham the average maximum temperature is 70° Fahr., at the Scilly Isles 64°.

Both influences affect the temperature in the winter also, but then the moisture-laden winds from the Atlantic play a more prominent part. They prevent the temperature from sinking as low as it might otherwise do. Here again the west has the advantage, for whilst Nottingham has an average minimum in January of 31° Fahr., the Scilly Isles have 42°. Thus at Nottingham the annual range is 39° Fahr., while that for the Scilly Isles is only 22°.

These facts illustrate the differences between a "continental climate" and an "insular climate." The latter is characterised by equableness and dampness, the former by cold winters and unusually warm summers. As compared with the climate of the Scilly Isles that of Nottinghamshire is continental. As compared with that of places of the same latitude in central and eastern Europe it is, in common with the climate of the whole of the British Isles, insular. The average temperature of our country is much higher than its latitude would lead us to expect. This is due to the fact that the prevalent south-westerly winds not only come from warmer regions themselves but also cause that drift of warm surface waters of the Atlantic to our shores commonly called the Gulf Stream. These combine to produce our mild winters.

The amount of rainfall varies considerably from day to day. In 1908 in Nottingham there were 182 days on which it was less than one-hundredth of an inch. The heaviest rainfall, over six-tenths of an inch, was on the 25th of March. The total fall for the year was 22·7 in., which is equivalent to a downpour of 25,086,152 tons of water upon the city.

The average rainfall during the last forty-two years has been 24·9 inches. During that period the wettest year was 1872 with a fall of nearly 36 inches, and the driest year 1887, with less than 16 inches.

The accompanying map, prepared by Mr Mellish of Hodsock, shows the average distribution of rainfall in this county for a period of thirty years. From this it will be seen that the rainiest district is the high ground west of

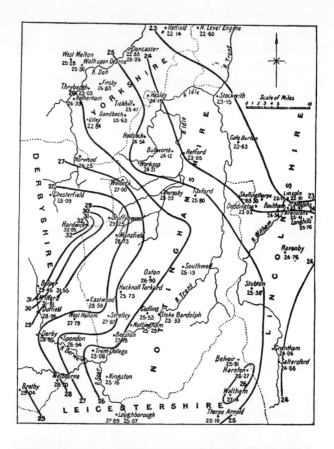

Map showing Distribution of Rainfall in Nottinghamshire
and neighbouring Counties

Robin Hood's Hills. It is interesting to note that this supplies the sources of many of the largest streams in the county. The driest district is the lowest ground in the extreme north of the county.

From the general rainfall map of England it will be seen that this county is much drier than counties of the same latitude west of the Pennines. Much rain comes with the cyclones from off the Atlantic. As the air laden with moisture from its passage over the ocean is forced up by the Pennines it expands and cools and consequently discharges much of its moisture there. Passing over the hills the air then descends to the low ground, including Nottinghamshire, on the east. The reverse now happens. The greater part of this county is therefore in the rain-shadow of the Pennines.

What these are to our county the moorland of Devon and Cornwall, the Welsh mountains, or the fells of Cumberland and Westmorland are to the rest of England. This accounts for the fact, well shown on the map, that the heaviest rainfall is in the west, and that it decreases steadily until the least fall is reached on our eastern shores. Thus in 1906, the maximum rainfall for the year occurred at Glaslyn in the Snowdon district, where 205 inches of rain fell; and the lowest was at Boyton in Suffolk, with a record of just under 20 inches. These western highlands, therefore, may not inaptly be compared to an umbrella, sheltering the country further eastward from the rain.

The rainiest month in the year is October, though July and August run it very close. In fact our county is characterised by its summer rains, as about 10 inches out of the annual 25 fall during the months of May, June, July, and August.

The climate of Nottinghamshire may be briefly summarised in the following terms:—the prevailing winds are westerly and south-westerly, the amount of sunshine is moderate, the temperatures comparatively extreme, and the rainfall low with a large fall in the summer. It must be remembered, however, that strictly local factors may exist. An open country or a sheltering range of hills, a southerly or a northerly aspect, a sandy or a cold clay soil will produce striking differences in the climate of places distant only a few miles one from the other.

11. People—Origin, Race, Population.

In the chapter on Natural History it was seen that the animals and plants of Britain came originally from the Continent. The same is true of its people.

How long man has existed on the earth no one can say. No even approximate guess can be hazarded, and it is not likely that our lack of knowledge in this respect will ever be enlightened. But from recent researches in Crete it seems reasonable to suppose that a high state of civilisation was in existence there, and no doubt in other parts of the Mediterranean basin, perhaps as much as, or even more than, 10,000 years ago. This civilisation never reached our shores, but in times far more remote than these man was existent in Britain. Almost all traces of him, save of the implements he made—chipped flints for hunting and warfare—have vanished, but we know that he was a contemporary of the mammoth, woolly rhinoceros, and cave bear when these animals inhabited our land, that he was a hunter pure and simple, with no domestic animals and doing little or nothing in the way of agriculture, and that he had not learnt to grind and polish his weapons. He is known as Palaeolithic man—man of the Old Stone Age.

How long a period intervened in Britain between him and his successor, Neolithic man, or man of the New Stone Age, it is impossible to say, but most authorities agree that the gap is very great. When our land became again peopled it was by a race probably Iberian, from south-west Europe, which, though still ignorant of the use of metals, had attained a certain degree of civilisation, possessing the domestic animals, raising crops, and knowing the arts of spinning and weaving. The Neolithic Age no doubt lasted for a very prolonged period.

How long this Iberian race had been settled in our land before the advent of the Celtic races we do not know. These latter came in two separate invasions and introduced the Age of Bronze. The Goidels or Gaels overran the country, driving their predecessors into the hills and fastnesses of Wales. Later the Brythons or Britons came, pushing the Gaels before them to the north; and at the beginning of our era the Romans appeared, bringing with them their legionaries drawn from the various countries of their conquest.

The Teutonic invaders of our island entered by the rivers which open on to the east coast. The Jutes and

Saxons settled in the east and south of England; the Angles and Danes in the midlands and the north.

The Angles seem to have travelled westwards from Lincolnshire along the old Roman Fosseway. From this they wandered into our county and dotted those parts not covered by marsh with farmsteads. The sites of their settlements are still indicated by the suffixes *ton* and *ham* in so many village names, as Ruddington, Hoveringham. The forest did not daunt them. They felled the trees and made clearings, the memory of which is still preserved in such names as Mansfield and Farnsfield.

After the Angles had become established the Danes came. Some of them entered Yorkshire by the Humber and thence worked southwards to Nottinghamshire. The grouping of such names as Serlby, Ranby, Scrooby in the north-west of the county is an indication of their complete settlement there. Gunthorpe, Knapthorpe and other like names may have a similar significance.

Last of all came the Normans. They also were Teutonic, but belonged to a branch of the race which had settled first in the North of France.

The population of Nottinghamshire is therefore mainly Teutonic, and might perhaps be roughly described as Anglo-Danish.

The Teutonic tribes which invaded Britain spoke different dialects. That spoken by those who settled in the Midlands became in course of time the English language. Hence it is that in this county the native dialect does not differ from English as much as does the dialect of the men of Yorkshire or Wiltshire.

The mixed nature of our ancestry has been well summarised by Professor A. H. Keane. "Britain," he says, "has been successively occupied by a great number of peoples—primitive man in the Old Stone Age; Picts, and perhaps others associated with the dolmens and other megalithic monuments, in the New Stone Age; tribes of Keltic speech, commonly called Kelts, in the Bronze Period, possibly as early as 2000 B.C.; Belgae or proto-Teutons somewhat later; Romans and their legionaries of diverse origins about the New Era; early and later Frisians, Saxons, Angles and others of Teutonic speech, say between 300 and 500 A.D.; Scandinavians, chiefly Danes and Norwegians, of kindred speech, eighth to tenth century; Normans, mainly Norsemen Romanised in speech, eleventh century; with sporadic arrivals from the mainland down to the present time."

There is no record from which we may gather the exact number of people in Nottinghamshire in early times. The first census was taken in 1801, when the population of the county was 140,350. By 1901 it had increased to 514,578 persons. Half of these were born in the county. The others were from every county in England and Wales as well as from other parts of the world. This fact illustrates the rapidity with which the population of the country is becoming mixed.

The average number of persons per square mile in 1901 was 610. The corresponding number for England and Wales was 558; for the most crowded county—Lancashire—2347; for the least crowded—Westmorland—82; for London, 38,774.

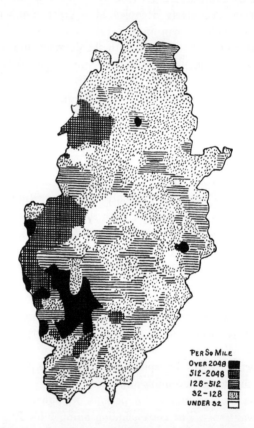

PER SQ MILE
OVER 2048
512-2048
128-512
32-128
UNDER 32

Map showing Distribution of Population in Nottinghamshire
in 1901

In Nottinghamshire the population was most concentrated on the west, especially along the Leen and Erewash valleys. The most thinly populated parishes were Welbeck, Rufford, and Wiverton Hall, each of which had less than twenty-five people to the square mile.

During the nineteenth century the population increased 397 per cent. in England and Wales and 400 per cent. in Nottinghamshire. In this county the greatest increase took place in the west. This was due to the exploiting of the underlying coalfield and the consequent growth of manufacturing industries. In the east, with the exception of Retford and Newark, the population, which is largely agricultural, was either stationary or had decreased. The decrease was most marked in that extensive piece of country between Rufford, Tuxford, and Newark. In some parishes the decrease amounted to as much as 70 per cent.

If the population of the whole county has continued to increase since 1901 at the same rate as it did from 1891–1901 it will amount in 1911 to 593,823, but there are indications that the number will be greater still.

12.   Agriculture—Crops and Stock.

Compared with many other parts of England and Wales Nottinghamshire is highly cultivated. In common with the rest of the country the area under cultivation, as recorded in the Board of Agriculture's returns, has steadily decreased during the last fifteen years. This is no doubt largely due to the increase in the number of gardens and of holdings less than one acre in extent of which no account is taken in the returns. These are a considerable item, for instance, in Nottingham, where extensive areas are given over wholly to gardens, which are said to number nearly six thousand.

Only five other counties exceed Nottinghamshire in the proportion of woodland. The area of woodland in 1905 was 28,540 acres, or five per cent. of the county. For England and Wales it is only four per cent. For many continental countries it is over 20 per cent. Since 1891 the area in our county has been increased by 2721 acres. This has not been done for economic purposes but for the improvement of the landscape or for the preservation of game.

In Nottinghamshire 42 per cent. of the area is occupied by arable land and 40 per cent. by pasture. Its general character may therefore be described as half corn-growing and half pastoral. Corn crops are grown in all parts of the county.

Wheat does best on the heavy clay lands of the Keuper and is favoured by the dry climate and warm summers. In 1907 there were 33,795 acres or six per cent. of the county under wheat. This is a smaller proportion than in corn-growing Lincolnshire with its nine per cent. of wheat land, and larger than in pastoral Leicestershire which had only four per cent. This difference is accentuated by the productiveness of the soils, for in Nottinghamshire each acre produces on the average 29 bushels whereas in Lincolnshire it produces as much as 34 bushels.

For some years the area devoted to wheat-growing
has declined all over Great Britain.   In this county in
1907 there were 4000 acres less than in 1906 and 30,000
less than in 1880.   This is associated with the low price
consequent upon the importation of large quantities of
foreign wheats.   Nevertheless it is still necessary to mix
English with foreign wheats in order to produce the best
qualities of flour.

The area devoted to the growth of barley has also
shown a similar but not so strongly-marked decline.   On
the other hand oats have been cultivated more and more.
The area so used advanced from 23,452 acres in 1880 to
39,767 acres in 1907.   The advance between 1906 and
1907 was over 4000 acres.

In 1907 the other corn crops—rye, beans, and peas—
occupied 2894, 5262, and 4750 acres respectively.

Corn crops are not grown upon the same ground
year after year, otherwise the quality of the crops would
decline.   In past years much of the land was left to
rest, i.e. to lie fallow.   Thus in 1880 as many as 24,332
acres lay fallow.   But in 1907 only a little over 7000 acres,
chiefly very heavy clay land, were thus left to produce
nothing.   In these days arable land is used for growing
other crops besides corn.   Some of these, such as clover,
vetches, etc., enrich the soil; others draw their nutriment
from different depths; whilst turnips, swedes, and man-
golds are so widely spaced that the farmer is able to hoe
and to clean the soil from weeds.   Thus whilst the soil
is being prepared for the growth of corn it is also being
used to produce winter fodder for cattle.   In 1907

28,523 acres were devoted to turnips and swedes, 51,348 acres to clover, sainfoin, and temporary grass, and 19,135 to other crops.

The chief areas for permanent pasture are along the sides of the Trent and Soar and in the Vale of Belvoir. The Trent-side pastures are used mainly for fattening

Stilton Cheese Making
(Midland Agricultural and Dairy College)

cattle. The rich pastures of the other two areas are devoted mainly to dairy farming. Sheep are reared all over the county ; least of all in the dairying district, most in the sandy lands of the Bunter, the dryness of which is less liable to cause diseases of the foot and liver. The number of sheep in Nottinghamshire is about half the

number of those in Leicestershire, and has decreased by
one-half in twenty-five years.

Within the county Cropwell is noted for sheep, Ruddington for cattle, Carlton-on-Trent for shire horses,
Edwinstowe for hunters, Colston Basset for pigs.

One hundred years ago the country around Tuxford
was famous for pigeons. It is recorded that "seven

Butter Making
(Midland Agricultural and Dairy College)

hundred dozen were sold on one market day at Tuxford
to a higler from Huntingdonshire."

No account of the agriculture of the county would
be complete without a reference to the Agricultural
and Dairy College at Kingston. This institution was
established by the County Councils of Nottinghamshire,

Derbyshire, Leicestershire, and Lindsey. It possesses a highly qualified staff, a thorough equipment, spacious premises, and indeed all the requirements for giving efficient instruction, theoretical and practical, in all branches of farming.

Concerning special cultivations there is little to be said. "Skegs," a poor quality of oats said to be peculiar to the county, were grown on very poor land and "reckoned a sweet food." Liquorice was at one time grown near Worksop, and weld or dyer's weed, *Reseda luteola*, used to be cultivated in large quantities south of Scrooby for making a yellow dye.

At the commencement of the nineteenth century hops were grown on the land between Retford and Tuxford and sold at annual Hop Fairs at these places. They were also grown around Southwell. The plantations were situated chiefly in valleys and wet situations. These hops were known in the trade as "North Clay Hops" and were considered much stronger than Kentish hops. At that time an area of no less than 11,000 acres was devoted to this cultivation. In 1880 it had dwindled to 29 acres. Now they are not grown at all.

13. Industries.

From the two preceding chapters it will be gathered that agriculture is the main industry for the greater part of the county.

North of the Trent and west of the line joining Nottingham to Mansfield there is a dense population engaged in mining and manufacturing industries. The manufacturing population is concentrated mainly in the city and suburbs of Nottingham, for it is a great commercial convenience to have the factories close together. The mining population is scattered about in straggling towns and villages, for of necessity the mines must be far apart. The importance of coal-mining in the county may be judged by the fact that in 1908 it employed close upon 37,000 men. But more will be said about it under the heading of Mines and Minerals.

The staple manufactures of the county are lace and hosiery.

Nottingham is the metropolis of the lace-making industry. Of the 25,000 people so engaged the great majority are women, and more than four-fifths are in this city. The industry has many branches. At the outset it calls for highly skilled and trained designers, for the sale of the lace depends largely upon the attractiveness of its design. One or several machines are set to manufacture one design until it has gone out of fashion. Sometimes the "run" is only for a few weeks. Occasionally it goes on for several years.

A fancy lace does not come from the machine in single narrow strips as it is sold. A great many strips are made at once side by side. Their edges are held together by long threads running their whole length and they thus form a great sheet—a "piece" as it is termed—many yards long and several wide. The width of the piece, and therefore the number of strips in it, depends upon the length of the machine. In the earliest this was only 18 inches. In the newest, driven by steam or electric motor, it is as much as seven or eight yards.

Whilst the lace is in the piece it is bleached, dyed, and finished. After this the strips have still to be set free one from another by drawing out the long threads. There are also numerous threads covering the lace which must be clipped away. Sometimes clipping frames are

Typical Nottingham Lace

used, but both pieces of work are done mainly by hand and find employment for thousands of people in their own homes. It is a familiar sight in the side streets of Nottingham to see women sitting at the open door on a warm summer's day drawing, clipping, and scalloping lace.

The main divisions of the lace-trade are edgings, insertions, curtains and nets, plain nets, and warp laces.

From Nottingham and the neighbourhood lace is sent
to all parts of the world, more especially to the United
States and South America. Large shipments are also
made to all parts of Europe and the British colonies.
The lace industry is the child of the hosiery industry, but
in Nottingham the child has quite outgrown its mother.

Plain Net and Curtain Machines in course of erection

The hosiery industry is not so concentrated as the
lace. About 6000 hands are employed by it in the city,
and over 5000 in the neighbouring towns and villages,
such as Mansfield, Arnold, Beeston, Hucknall Torkard,
Ruddington, etc.

Until the sixteenth century every family made its
own hose. In 1589 the Rev. William Lee of Calverton,
near to Nottingham, invented a machine for making

stockings. The story of his struggles should be familiar to every boy and girl. This machine came to be called the stocking-frame and could be driven and worked by one man. In the early part of the seventeenth century frame-knitters were proud of their profession and wore a silver knitting-needle as their badge. As the demand for machine-made stockings increased the industry spread to most of the villages of this county and the adjoining parts of Derbyshire and Leicestershire. When water-, and afterwards, steam-power were introduced it became possible to make and work larger and more complex machines for the manufacture not only of stockings but of many other kinds of hosiery. Naturally the hand-frame workers of the villages were placed at a dis-advantage and the industry became concentrated in the towns, especially Leicester, Derby, and Nottingham. Throughout the neighbouring parts of the three counties the wide windows in the top stories of many of the houses tell of the times when this village industry flourished un-rivalled. Even to-day in some villages, e.g. Calverton, Burton Joyce, etc., the zz*it*, zz*it*, zz*it*, of the stocking-frames announces the fact that they are still used to a limited extent in the manufacture of various kinds of hosiery.

For many centuries Nottingham was noted for its smiths. Evidence of their former importance is still forthcoming in the names of many of the streets. The iron for their work was obtained from Pleasley and Bulwell, where the ore from Derbyshire was smelted with the wood from Sherwood Forest. In those days

Old Stocking Frame

the smiths were engaged in the making of parts of harness and of agricultural implements. In the seventeenth and eighteenth centuries their skill was turned to the making and improving of the stocking-frame. To-day it shows itself in the complicated machines required for producing lace and hosiery, and to a less extent, in many other kinds of machinery. The most modern development of this class of work is seen in the growth of the cycle and motor industries.

Side by side with the smithy work flourished the leather-tanning industry. This owed its importance largely to the natural advantage of an abundant supply of bark from the neighbouring forest. This industry is still carried on, but not to the same extent.

Though it does not employ many people the colour-printing done in Nottingham nevertheless stands in the front rank of that class of work in the United Kingdom.

It has been already noticed that the lace and hosiery industries employ a great many women. There are several other important industries which do likewise, e.g. tobacco and cigar-making, laundry work, the "making-up trade" or the making of blouses, skirts, aprons, costumes, and the like.

The importance of Nottingham tends to overshadow the smaller industrial centres in the county. Foremost among these is Newark, which has long been an important brewing, malting, flour-milling, and agricultural-engineering town. Other centres are Mansfield, Sutton-in-Ashfield, Hucknall Torkard, Southwell, Retford, Worksop, and West Stockwith.

14.   Mines and Minerals.

Two important minerals are raised in our county, viz. coal and gypsum.

Coal has been used in Nottingham for many centuries. On one occasion Henry III had to leave his Queen in

Open Working of Coal Seam, Erewash Valley

this town for a time, but she soon departed because she could not endure the smoke from the coal. In 1641 "stone coal" was got at Wollaton, Strelley, and Bramcote; and was conveyed to Nottingham and sent thence down the Trent to Newark, Gainsborough, and other places.

At first the coal was quarried in open workings

situated on the outcrop of the seams. As the workings were carried forward the overlying rock became very thick, and rain-water ran down into the quarries, thus making it more difficult and expensive to get the coal. The miners then found it better to "mine" rather than quarry the coal, and so they went further east and sank shallow pit shafts to the coal-seams. The coal was then extracted from between roof and floor and the mine kept open by the erection of wood props. Because the seams sloped further into the earth those made later were gradually sunk more deeply. In the eighteenth century such mines existed at Bilborough, Brinsley, Eastwood, Teversall, and Wollaton. At the first-named place the coal was worked at a depth of 300 feet. In the middle of the nineteenth century engineers began to sink mines through the newer rocks, and since then many others have been sunk further and further towards the east, more especially in the Leen valley. Generally speaking the further east a mine is situated the deeper must it go. The most easterly ones at present are the Gedling and Manton pits. In the former coal is worked at a depth of 1377 feet. In the recent boring at Oxton, which is still further towards the east, the same seam was reached at a depth of 2030 feet. As the sinking of a shaft is very expensive, fewer mines will be put down in this part of the coalfield than in the west. On the other hand a greater area will be worked underground from one shaft. It may be expected, therefore, that the collieries of the future will be more widely spaced. The mode of working is called "Longwall." Notting-

hamshire miners are very proud of it and claim that it
is peculiarly their own. It is now being adopted in all
the other coalfields of England, and all over the world
"Nottinghamshire Longwall" is spoken of with great
admiration. By this system all the coal is extracted by
workings which day by day advance further outwards.
Thus after a life of 30 or 40 years the working places

Gedling Colliery

underground may be two or even three miles distant from
the shafts. From the bottom of the shaft several spacious
passages called "gates" radiate. These branch and re-
branch until they reach the "working face" at many
points. At the "face" the seam is seen to be several feet
thick and is overlain and underlain with other rocks. As
the collier removes the coal he places it in little trucks
or "tubs." Any waste or rock he has removed in

winning the coal he builds up in the space behind. The loaded tubs are pushed on rails laid close to the face until they reach a gate. In the gates they are drawn by horses to the " haulage-planes " where they are attached to wire-rope cables and are transported by steam or electric power to the bottom of the shafts. Thence they are raised to the surface by powerful steam engines often at a speed of 50 to 60 feet per second.

Large modern collieries are capable of raising upwards of 3000 tons of coal during a working shift of eight hours. A single mine will find employment underground for considerably over 1000 men. Large volumes of air are therefore required for them as well as for removing the dangerous firedamp which escapes from the coal seams. There are two shafts to each mine, the wind goes down one and up the other, being drawn by a revolving fan which may circulate a current of 200,000 or 300,000 cubic feet of air per minute through the mine.

In this coalfield the most valuable seams are known by the following names :—Top Hard, Deep Soft, Deep Hard, and Furnace. The first of these, which is 4 to 5 feet thick, is the most important and is identical with the famous Barnsley coal of Yorkshire, though only half as thick. The distance in depth between the first and last is about 900 feet. In 1870 the output of coal in this county was only two million tons. There are now over 50 separate collieries, the output of which during 1908 amounted to 11,044,214 tons. This is about six millions less than in Derbyshire, one-third the output of Yorkshire, and one-seventeenth of that for the whole of England.

The Nottinghamshire coal-mining industry, however, is as yet only in its infancy. Its coalfield is part of a larger one known as the York, Derby, and Nottinghamshire coalfield. The Royal Coal Commission report in 1905 shows that the resources of this larger field are as great as those of South Wales and Monmouthshire, and that there is enough available coal in it to last another 500 years at the rate of production for the year 1903. This rate however increases year by year.

Gypsum is, of course, not such an important mineral as coal, but Nottinghamshire is the most important gypsum-producing county. In 1908 the total output for the county was 29,685 tons, which represented one-third of that for the whole of the United Kingdom. Gypsum is found in the Keuper marls and is obtained usually by running tunnels into the hillside for several hundred yards. A gypsum layer varies in thickness from 15 feet to a mere film. It may be continuous or be made up of ball-like masses. The purest mineral is roasted and ground up to a powder as fine as flour. This is plaster of Paris. It is also used as alabaster for making ornaments and ornamental building-stone. One form has a beautifully silky appearance and is called "satin spar." This used to be found in abundance close to East Bridgford and is believed to have given origin to the name of a Roman station close at hand, viz. Margidunum, which probably means "Pearl Hill." The presence of gypsum in the waters in the vicinity of Newark makes them hard but peculiarly suitable for brewing.

Other less special products of the ground are obtained.

The clays of the Keuper marl yield a medium quality of bricks. On Mapperley Hills extensive brickyards have existed for many years (see p. 26). These have supplied much of the building material for the neighbouring city. Hence the saying that "Nottingham was once on Mapperley Hills." The Permian marls produce a better quality of bricks and tiles, and are also used for making coarse pottery.

The Magnesian Limestone yields an excellent building-stone, especially around Mansfield, where it contains much sandy material. It was from this formation that the stone was obtained for building Southwell Cathedral and the Houses of Parliament.

The limestone in the Lias, more particularly at Barnstone, is used for the manufacture of hydraulic cements.

15.   Water-Supply.

Water is not usually called a mineral but in this county it is a very important product of the underlying rocks. For man's well-being and for household purposes a good supply of it is just as essential as a good supply of coal. From the standpoints of sanitation and general health there is a growing feeling that even agricultural districts should no longer rely for their supply upon surface wells.

At present the needs of the city and many country places are amply satisfied from what we may describe as a gigantic subterranean reservoir in the Bunter sandstone. It has been already noticed that the outcrop of this rock is characterised by the fewness of the streams upon it. This is associated with the fact that here the rain water, instead of flowing off, soaks immediately into the ground. Much of it is taken up by vegetation or evaporates from the surface. It is probable that, at the most, only one-third percolates downwards. The outcrop of the sandstone covers an area of 183 square miles. The layer is of course thinnest along its western margin : elsewhere it varies in thickness up to as much as 750 feet in the north. The whole of this mass is saturated with water to within 20 feet of the surface on the eastern side. It is not difficult to sink an ordinary well to that depth. This fact accounts for the presence of a string of villages, viz. Arnold, Calverton, Farnsfield, Oxton, and Retford, along this edge of the Bunter.

It is natural to ask, "If the rock is so absorbent, why are not the streams which run over it absorbed?" There can be no doubt that in some cases the floor of the valley lies below the saturation level of the rock.

During the last forty years this reservoir has been drawn upon more and more by the sinking of deep wells. One of the largest of these is about 30 feet by 12 feet in cross section and goes down 150 feet. From the bottom tunnels are run into the surrounding rock. From the roofs and walls of the tunnels water pours like a concentrated thunder-shower and runs in a torrent along the floors to the bottom of the well. Thence it is raised to the surface by powerful pumps kept constantly at work,

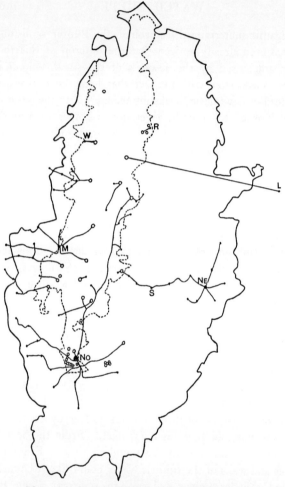

**Map showing the Distribution of Deep wells
in Nottinghamshire**

*(The wells practically all in the Bunter Sandstone, the outcrop of which
is indicated by the dotted line. The lines show the water mains
from the wells, and the dots the towns and villages which at present
derive their supply from deep wells in Bunter)*

some of them being capable of delivering between two and three millions of gallons per day. From the pumping station the water is forced up to reservoirs situated at a height sufficient to secure their being above the highest point which requires a supply.

The accompanying map shows the position of most of the deep wells in the Bunter and the places supplied by them. Thus though the soil which the rock produces is poor and the surface-water is limited yet it is a valuable asset not only to this county but also to the adjoining ones. Our forefathers avoided its outcrop as a place for settlements. We now go to it for the water required by the settlements they established.

It must not be supposed that this reservoir is inexhaustible. Several pumping stations now yield only half the supply they yielded at first. This is partly due to the fact that they have pumped the water away more quickly than it has percolated from above. Nevertheless it will always be a valuable and economical source of supply even though the increasing demands of a growing population may render it necessary to augment this from other counties in the years to come.

16. History of the County.

Let us now turn to the history of man in Nottinghamshire. The hills, valleys, and rocks which make up the county have a history of their own. They are the stage upon which man has acted. For ages generation has been followed by generation, but the hills, streams, climate, and natural history, which have been dealt with in earlier chapters, have altered very little.

The present generation is writing its history in books and papers. It is also making less perishable records in the form of railway lines, factories, and mines. This

Newark Castle and the Great Cross Roads

(*The Fosseway is seen to left of the picture, the Great North Road to right*)

chapter is concerned only with the written history of past generations. The succeeding ones attempt to give body and life to that history by the contemplation of their handiwork.

Nottinghamshire with its extensive forests on the west and swamps on the north and east offered no great attractions to the Romans. They had no stations of such

importance as Lincoln and Leicester within the bounds
of the future county. But it was traversed by the great
road, the Fosseway, which joined these two places. This
road, which enters the county near Collingham and leaves
it near Willoughby on the Wolds, was the probable route
by which the Angles invaded the county during the sixth
century. A great battle was fought between the East
Anglians and the Northumbrians in the year A.D. 617
on the Idle close to Retford.

The Angles soon realised the strategic importance of
what we now know as the Castle Rock at Nottingham,
which must have played an important part in the struggle
between the rival kingdoms of Mercia and Northumbria.
The first mention of this stronghold is made in connection with the coming of the Danes. These latest
invaders, having gained control over Northumbria, came
south from York and entered it in A.D. 868. Here they
wintered, but in the following year were attacked and
compelled to make a treaty with Ethelred and Alfred.
Six years later they returned, but this time they extended
their conquests far beyond Nottingham into western
Mercia. In A.D. 878 Alfred concluded the Peace of
Wedmore with them. By this it was agreed that they
should have control over the country north-east of Wat-
ling Street. This region was called the Danelaw. Here
they had several strongholds known as " The Five
Boroughs," of which Nottingham was one. In or about
A.D. 922 it was recaptured by the Saxon king of Wessex,
Edward the Elder. He built a bridge across the Trent
at this point and placed fortifications near each end.

Two years after the battle of Hastings William I came to Nottingham preparatory to subduing the north. Here he erected a castle which he entrusted to one of his followers, William Peveril, and thus gave permanence to his conquest in this region.

From that time until the Commonwealth Nottinghamshire was a constant resort of nearly all the sovereigns with

Market Place, Newark

the exception of the Tudors. Several factors account for this. It occupied a central position in the kingdom and lay upon the great trunk routes to the north. For the earlier kings especially, Sherwood Forest was a favourite hunting-ground and the castle a worthy residence.

Henry II took the castle from the Peverils, and made

Nottingham an object of his favour, granting a charter to its burgesses.

During the absence of Richard I from England this district became the centre of his brother John's treacherous activities. On his return Nottingham held out for John to the last and did not yield until the King began to carry the castle by storm. Both this town and Newark always remained true to John. The former received three charters from him and at the latter place he died.

After Queen Isabella and Roger Mortimer had contrived the murder of Edward II they established themselves in Nottingham Castle and sought to rule the country during the minority of Edward III. In 1330 this youth, then eighteen, proceeded to take over the reins of power. With several nobles he gained access to the castle by means of a secret passage now known as "Mortimer's Hole." Mortimer was arrested, sent to the Tower of London, and finally executed at Tyburn.

In 1349 Nottinghamshire in common with the whole country suffered from that scourge, the Black Death. The depopulation caused by this led to consequences of far-reaching influence, more particularly on the relationships of landowners to labourers and upon the discipline of the clergy.

During the Wars of the Roses the county was mainly on the side of the Yorkists and the castle was generally in their hands. Richard III made it his headquarters. It was from here that he started out on that journey which ended so fatally on Bosworth Field.

After the close of the wars the Yorkists used Lambert

Simnel as their tool, and having crowned him Edward VI in Dublin they determined to depose Henry VII. They landed their forces in Lancashire, marched first to York, and then across Nottinghamshire by Mansfield and Southwell to Fiskerton, where they forded the Trent and encamped on the opposite side of the river at East Stoke. Meanwhile the King had collected his army at Nottingham and held a council of war in the castle. The next day he marched to Newark, and on the following day, June 16, 1487, joined in battle with and defeated the rebels on Stoke Field.

On August 25, 1642, Charles I performed the opening act of the Civil War by erecting his standard on the present grounds of the Nottingham General Hospital. Here he called upon his subjects to rally round him, but received no very hearty response. Generally speaking the landowners and Newark sided with the Royalists; the people and Nottingham with the Roundheads. The two castles held similar strategic positions for the two parties. Both were keys to the north; the one for the Parliament, the other for the King. Though no great battle was fought in this county it witnessed the closing as well as the opening act of the war. In May, 1646, Charles surrendered himself to the Scottish Commissioners at the Saracen's Head Inn, Southwell. From thence he sent orders to Newark, the last of the Royalist strongholds, to yield.

17. Antiquities.

Long before man could write he could make. Hundreds of generations who spent their lives in this district around us left no written record. All that can be known about them must therefore be learned by the study of their handiwork.

Church Hole Cave, Creswell Crags

Perhaps it will be most instructive to start with the remains found at Creswell Crags. Here a small tributary of the Poulter forms two miles of the county boundary, which at one point passes through a picturesque ravine in a range of Magnesian Limestone hills (p. 7). In the cliffs on either side there are several caves, of which an important one, the Church Hole Cave, is in Nottinghamshire.

The first inspection of this cave showed that it had been used as a stable within quite recent times. Further digging in the earth covering the floor near the mouth revealed a bronze brooch, a bone awl, and ware made by Britons who had learnt the art from the Romans and had made their home here. Below this earth were several distinct layers, the uppermost of which contained charcoal— always a sure sign of man's presence—flint implements, a bone needle and an awl. The last two indicate that the original owners clothed themselves in skins which they pierced by means of the awl and then sewed together with the aid of the needle. In the middle layers less perfectly worked flint and bone implements were found together with others of a rougher type made from the quartzite pebbles which are so common in the Bunter. In the lower layers only implements of the last type were found. Side by side with these implements were found the bones of animals which became extinct ages ago in these islands, viz. the mammoth, reindeer, woolly rhinoceros, bear, lion, and hyaena.

The history of man in this county, as far back as we are at present able to read it, begins with those rough quartzite implements in the lower layers. These, together with the animal remains, show that he lived during the Old Stone or Palaeolithic age. Within the period of time represented by the middle and upper layers he learnt to use flint and bone, and became a much more accomplished workman. He even showed some rudiments of art, and in a cave across the Poulter was found a bone with the drawing of a horse's head roughly scratched upon it.

From the time when the cave was occupied by Palaeolithic man to the time when it was used as a home by post-Roman Britons a vast period elapsed, during which another race hunted in the woods and on the open wolds, or herded their cattle and tilled the soil. This was Neolithic man, whose more perfect stone implements, many of them smoothed and polished, have been found chiefly in the south of the county. Recently some were discovered in the base of the alluvium of the Cocker Beck.

The Neolithic race was in its turn, as we have seen, gradually conquered and displaced by another, which continued for a time to use stone but had learnt how to use bronze. Their earliest bronze implements were merely imitations of the rudely-shaped stone ones ; but in time they gave place to beautiful axe-heads, spear-heads, and swords. A good collection of these was found in 1860 in some excavations in Great Freeman Street, Nottingham. When the Romans came, not so much to colonise as to exploit the mineral wealth of this country, this was the race which occupied the land.

To the Romans Nottinghamshire seems to have offered very few attractions, consequently their remains are comparatively few. Roman pottery and coins and portions of houses have been found at Brough, Margidunum (East Bridgford), and Mansfield Woodhouse ; and the foundations of a bridge across the Trent at Cromwell.

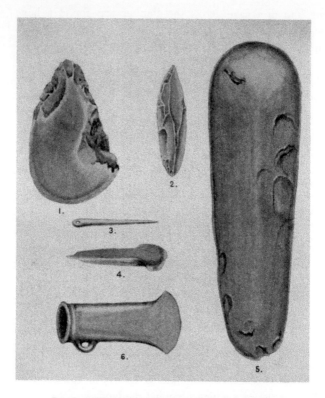

Implements of the Stone and Bronze Ages

1. Quartzite Palaeolith (Robin Hood Caves) ⎱
2. Flint   „   „   „   „   ⎰ after Boyd Dawkins.
3. Bone Needle⎱ Church Hole Cave
4. „ Awl ⎰
5. Neolithic Hatchet. Charlotte St, Nottingham⎱ From specimens in Museum
6. Bronze Palstave. Gt Freeman St,   „   ⎰ Nat. Hist., Nottingham.

A pavement on the bed of the river at Littleborough still marks the point at which the Roman road from Lincoln to Doncaster and York crossed the Trent. Portions of other roads exist here and there, but the most enduring monument of Roman skill and industry is the Fosseway.

The departure of the Romans was followed by a

Hemlock Stone

series of Anglian invasions. Here again written history has been supplemented by the discovery of handiwork. In 1842 the site of an Anglian burial ground was found at Holme Pierrepont. Two years later a similar site with urns containing the ashes of cremated dead was found at Kingston-on-Soar. In 1893 the grave of a warrior with sword and spear was discovered at Aslockton. Several brooches of a type peculiar to this part of England have also been dug up.

No prehistoric stone monuments occur in the county. The nearest approach to anything of the kind are the Hemlock Stone near Bramcote and similar stones at Blidworth. These, however, were formed by natural processes. They resisted the destroying action of the rain and frost more effectively than the surrounding soil

"Cranmer's Mound," Aslockton

and rock, and whilst these have been worn away they have remained. They impress men even now, and there are indications that they were held in reverence by early man.

Mounds or tumuli which were probably ancient burial places occur at Blyth, Blidworth, and Oxton.

Practically all the other earthworks of the county were originally used for military purposes. At Oxton the encampment known as Oldox is roughly oval in shape. It has several tiers of ramparts and a circuitous entrance, both of which features are characteristic of Celtic handiwork. Rectangular encampments, probably made by Romans, exist west of Oldox and on Cockpit Hill. Near to Laxton there is a well-preserved hill fortress with a mount and two courts. This is almost certainly of early Norman origin. Similar works occur at Annesley, Aslockton, and Egmanton, and formed the original castle at Nottingham. The Sconce hills and other earthworks around Newark were made during the Civil War.

18. Architecture—(a) Ecclesiastical.

A preliminary word on the various styles of English architecture is necessary before we consider the churches and other important buildings of our county.

Pre-Norman or, as it is usually, though with no great certainty termed, Saxon building in England, was the work of early craftsmen with an imperfect knowledge of stone construction, who commonly used rough rubble walls, no buttresses, small semicircular or triangular arches, and square towers with what is termed "long-and-short work" at the quoins or corners. It survives almost solely in portions of small churches.

The Norman Conquest started a widespread building of massive churches and castles in the continental style called Romanesque, which in England has got the name of "Norman." They had walls of great thickness, semi-

Norman Porch, Balderton Church

circular vaults, round-headed doors and windows, and lofty square towers.

From 1150 to 1200 the building became lighter, the arches pointed, and there was perfected the science of vaulting, by which the weight is brought upon piers and buttresses. This method of building, the "Gothic," originated from the endeavour to cover the widest and loftiest areas with the greatest economy of stone. The first English Gothic, called "Early English," from about 1180 to 1250, is characterised by slender piers (commonly of marble), lofty pointed vaults, and long, narrow, lancet-headed windows. After 1250 the windows became broader, divided up, and ornamented by patterns of tracery, while in the vault the ribs were multiplied. The greatest elegance of English Gothic was reached from 1260 to 1290, at which date English sculpture was at its highest, and art in painting, coloured glass making, and general craftsmanship at its zenith.

After 1300 the structure of stone buildings began to be overlaid with ornament, the window tracery and vault ribs were of intricate patterns, the pinnacles and spires loaded with crocket and ornament. This later style is known as "Decorated," and came to an end with the Black Death, which stopped all building for a time.

With the changed conditions of life the type of building changed. With curious uniformity and quickness the style called "Perpendicular"—which is unknown abroad—developed after 1360 in all parts of England and lasted with scarcely any change up to 1520. As its name implies, it is characterised by the perpendicular arrangement of the tracery and panels on walls and in windows, and it is also distinguished by the flattened arches and the square arrangement of the mouldings over them, by the elaborate vault-traceries (especially fan-vaulting), and by the use of flat roofs and towers without spires.

The mediaeval styles in England ended with the dissolution of the monasteries (1530–1540), for the Reformation checked the building of churches. There succeeded the building of manor-houses, in which the style called "Tudor" arose—distinguished by flat-headed windows, level ceilings, and panelled rooms. The ornaments of classic style were introduced under the influences of Renaissance sculpture and distinguish the "Jacobean" style, so called after James I. About this time the professional architect arose. Hitherto, building had been entirely in the hands of the builder and the craftsman.

When the early builders wanted a larger place of worship they did not pull down all the work of their forefathers but added to or altered it. Consequently many of their buildings exhibit the workmanship of several periods. Amongst these Southwell Cathedral stands out pre-eminently.

The nave and transepts were built about 1110 in typical Norman style. The semicircular arches, sturdy round pillars, and massive walls were intended to carry a stone roof, which however was never completed. The dark recesses of the triforium and the brightness of the clerestory windows above produced broad contrasts of light and shade.

The Choir was built between 1230 and 1250 during

Southwell Cathedral

the Early English period. The builders had now learnt to appreciate and use pointed arches. The loftiness of these encroached upon the space given to the dark triforium, but finer light and shade effects were now produced by deep mouldings upon the arches and capitals. The windows were made lancet-shaped, and later several were grouped together, thus foreshadowing those of the next period.

The Chapter House was added from 1285 to 1300 during the Decorated period. Stone carving now reached its sublimest height. The great weight of the stone roof was now thrown entirely upon deep buttresses. Thus it became possible to make spacious windows by bringing together several lancets and opening out the space above with a geometrical pattern. This gave ample room for the display of the beautiful stained glass which was rapidly becoming the pride of the age.

In the latter part of the fourteenth century the builders of the Perpendicular period spoiled the grandeur of the west front by inserting a great window with mullions rising straight from the bottom to the top. Its lack of harmony with its Norman surroundings did not trouble them. They thought only of the glory which would flood the nave as the rays of the setting sun streamed through the broad expanse of beautifully stained glass.

Compared with Northamptonshire or even with its neighbour Lincolnshire this county is not rich in parish church architecture. Nevertheless there are examples of more than ordinary interest. A large portion of Blyth church is older than Southwell Cathedral. Newark parish

The Nave, Southwell Cathedral

church combines all the pointed Gothic styles, and its spire is one of the finest in the kingdom. St Mary's, Nottingham, is very largely in the Perpendicular style, and its numerous windows have caused it to be compared with a lantern. Thurgarton church is a gem of Early English work. At Newstead Abbey the west front of the original priory

St. Mary's Church, Nottingham

church stands out as a beautiful though ruined example of Early English and Decorated work.

A study of several churches in any part of the county will furnish illustrations of all the styles. Only a few can be mentioned. The chancel arch at Carlton-in-Lindrick supplies us with an example of Saxon work. The Norman style is illustrated by beautiful arcades at the same place and at South Collingham and South Scarle, also by the chancel arch at Everton and doorways at Balderton, Carlton-on-Trent, Laneham, Teversall, and Woodborough. Early English naves exist at Car Colston, Lowdham, South Leverton, and Everton. Spires, which on the whole are characteristic of the Midland counties, are not so numerous here as in Leicestershire. Most of them belong to the Decorated period, e.g. Bingham,

Holme Church

Mansfield Woodhouse, Newark, and St Peter's, Nottingham. On the other hand many of the square towers were erected during the Perpendicular period, as Rolleston, Hawton, Keyworth, and Shelford.

Who were the architects of these buildings? It may be said that there were none in the sense we understand the term now. The men who built these magnificent cathedrals of our land, which our most practised architects cannot equal, were little more than practical builders. From the earliest times Christianity has had men who have devoted their lives to preaching its messages to and organising its worship for the public. To them we owe the Collegiate (now Cathedral) Church at Southwell, and practically all the parish churches.

Distinct from such men were the monks, who lived a life of seclusion in monasteries and ordered their lives according to severe rules. In the sixth century Benedict reformed and reorganised the system in Western Europe. Thus arose the Benedictine Order, which had a large priory in this county at Blyth. In the tenth and eleventh centuries new Orders were founded with similar and often stricter rules. The great priory at Lenton belonged to the Cluniacs, whose parent institution was at Cluny. The Carthusians, whose order was founded at La Chartreuse near Grenoble, had an establishment at Beauvale. The Cistercians, who had an abbey at Rufford, came from Citeaux (Cistertium) in Burgundy.

Some of the first class also adopted a monastic life. Their rules were not as severe as those of the monks, and they followed the teachings of St Augustine. These were therefore called the Augustine or Austin Canons. They were great architects and we possess examples of their handiwork in Newstead Abbey (p. 96), and in the priory churches of Worksop and Thurgarton. The Praemonstratensians, who were an offshoot from the Austin Canons, had an abbey at Welbeck. The Gilbertine Canons had a priory at Mattersey.

In the reign of Henry VIII all the monasteries were suppressed, the monks dispersed, their funds used for other purposes, and their lands given away or sold. In many cases the buildings became mere quarries, and consequently much of the stonework of these old architects was ruthlessly destroyed. Their greatest work however can never be thus obliterated. They have made a lasting impression on the religion and literature of our land. Moreover they were the chief agriculturists of the Middle Ages; they developed commerce and made England the chief wool-growing country of Europe.

19.   Architecture—(*b*) Military.

The traveller approaching Nottingham from the south or west is attracted by the sight of a stately mansion perched on a great rock which towers far above the tallest chimney-stacks. This building is quite modern, but its name—Nottingham Castle—is very ancient. Three centuries and a half ago the sight which would have met his eye would have been more like that shown in the picture of Nottingham Castle in the sixteenth century (p. 128). Then a typical mediaeval castle existed. Now all that remains of it are a few fragments of the foundations, and the barbican or fortified gateway shown in the front of the picture. This has been recently restored, but portions of the old stonework may still be seen peeping through the new covering. Formerly this gateway led into the outer bailey or courtyard, the site of which is now occupied by the lower grass-plots and flower-beds. On the opposite side of this yard was a bridge ending in a drawbridge across the moat, which still exists. Beyond was another gateway with a portcullis. To the right was the wall of the inner bailey or quadrangle, fortified with several bastions or round towers. That nearest the gateway had narrow openings through which archers might pour volleys of arrows upon an enemy who had gained the outer yard and was endeavouring to cross the bridge.

The inner bailey occupied the site of the present top lawn. On its north side was the great tower, which can be seen standing out against the sky in the background of the picture. This was erected by Edward IV and added to by Richard III. Its lower stages and spiral staircase still exist in neighbouring grounds. All the other parts of the castle were built before the time of Henry IV.

From the quadrangle an incline led up to the highest part of the rock. This is now occupied by the Art Museum, but then by the oldest part of the castle, the fortress built by William I in 1068. The entrance to this fortress had no portcullis. The space within was called the inner ward. From it a subterranean passage now known as Mortimer's Hole led down to the foot of the rock. On the north side was the keep, a very characteristic feature of Norman fortresses. It was usually a strong rectangular building which served as the last resort when all the other parts were carried by the enemy, and it almost invariably had a well.

Such then was Nottingham Castle, a favourite resort

of kings. Some idea of its importance may be gained by the fact that during its existence more Parliaments were held in it than at Westminster. It was at its best in the reign of Richard III. In common with other castles in the country it was systematically neglected by Henry VII and then began to fall into decay. At the close of the Civil Wars it was dismantled by Colonel Hutchinson. Some time after the Restoration it passed into the hands of the Duke of Newcastle, who preserved the outer gateway for use as a lodge but razed the remainder of the ruins to the ground. He then, at great cost, erected the present building as a residence. In 1831 all but the outer wall of this was destroyed by fire during the Reform Riots. In 1875 the site was leased to the Corporation of the City and the building was restored and converted into an Art Museum.

Newark also had a great castle, and fortunately considerable portions of it are still well preserved. It was commenced in 1130 by Bishop Alexander on the flat ground between the Devon and the Fosseway. The general plan seems to have been that of a keep on a large scale, rectangular in shape with a square tower at each corner. On the west it was protected by the river, on the other sides by a double moat. The entrance was on the north, and consisted of a barbican on the outer edge of the moat, a very strong gatehouse used as a keep on the inner edge, and a drawbridge between. The gatehouse, part of the adjoining wall, the lowest parts of the west wall, and the whole of the square south-west tower are all that remain of the Bishop's Castle. Traces of typical

Newark Castle

Norman archways and windows may still be seen. These together with the building material, which was oolite, help to distinguish the bishop's work from that of later date.

During the first quarter of the thirteenth century the castle was completed. Grey Lias limestone was used on the inside and red sandstone on the outside. The fine hexagonal tower at the north-west corner, a similar one in the middle of the west wall, and a Great Hall lighted by means of large pointed windows, were built at this time. In the fifteenth century the castle was used merely as a residence. To adapt it for this purpose the Great Hall was divided into two storeys having several rooms. Windows of late Perpendicular style and the beautiful oriel window which lighted the state room were also inserted.

Newark Castle was the focus of the Civil War, and soon after this it was allowed to fall into ruins.

The remnants of an old castle known as King John's Palace exist at Clipstone.

20. Architecture—(*c*) Domestic.

The architectural history of Newark Castle illustrates the early history of domestic architecture. In the twelfth century, when the building of the castle was started, might was right and the man who could not defend his property lost it. The lords of the land therefore built great strongholds for themselves and extended protection

Newstead Abbey

to the people around in return for service. Of the dwellings of the latter no examples have survived the ravages of time. The strongholds, on the other hand, have weathered the storms of centuries.

When the lord of Newark Castle visited it he lived in the keep-like gatehouse, whilst his serfs and servants were content to shelter themselves in flimsy wooden structures around the court.

In the thirteenth and fourteenth centuries local enmities died down and the nation became prosperous. During this time the Great Hall, 130 feet long and 22 feet wide, was built, with kitchens, chapel, etc. adjoining. Here everyone from the highest to the lowest dined and nearly everyone slept. Here also liberal hospitality was meted out to all comers.

In the fifteenth century serfs were no more, the need for armed retainers had gone and only the lord's family and domestic servants remained. These no longer lived together. The great hall was divided into small private apartments lighted with spacious windows. Thus the military stronghold with crowded garrison became a commodious family residence.

Newstead Abbey furnishes a different kind of history. Up to the sixteenth century it was the dwelling of a community of clergy. Then it was suddenly transferred to the hands of a lay family, and the chapel was subsequently allowed to fall into ruins. The bare cloisters, guest-chambers, refectory, and other apartments are now luxuriously furnished, and together with additions made from time to time, form a very beautiful residence.

The old archbishop's palace at Southwell was commenced about 1360 in the Decorated style and was completed with the addition of a great hall in the Perpendicular style about 1439. All except the latter fell into ruins. This is now incorporated in the recently-built bishop's palace.

Wollaton Hall

Wollaton Hall, the seat of Lord Middleton, is a complete and fine example of an Elizabethan mansion. It was commenced in 1580 and finished in 1588. The stone is said to have been carried by donkeys from Ancaster and paid for in coal. It was built at a time when the need for fortified dwellings had passed away, when the Gothic style was falling into disrepute and the interest in classic style was reviving. It therefore exhibits a combination of these two with features peculiar to the period.

Welbeck Abbey, the seat of the Duke of Portland, Clumber that of the Duke of Newcastle, Thoresby and Rufford Abbey those of Earl Manvers and Lord Savile respectively, were all built within comparatively recent times.

Clumber

Most of the old manor houses have long since disappeared and their sites are now known only by traces of the moats which formerly surrounded them. Hodsock Priory still occupies the original site and possesses a moat, bridge, and gatehouse. Wiverton Hall also has retained its ancient fortified gatehouse. Costock Manor was probably built in the sixteenth century.

At Brough remains of Roman dwellings with stone foundations and timbered walls have been found.

During mediaeval times most of the dwellings were timbered or frame houses. Some of these still exist at Newark. The Old White Hart inn is perhaps the oldest house in the county. The character of its ornamental plaster-work indicates that it was built in the fourteenth century. The Saracen's Head inn, where Jeanie Deans stayed, and the old house occupied by the Governor during the sieges of the Civil War, are later examples of timbered houses.

During the sixteenth century brick buildings became common and in the early part of the next century gained the ascendancy. The first brick house in Nottingham was built in 1615 and was demolished a few years ago. Other old ones exist in Narrow Marsh and Bulwell. The finest example of old brickwork is in High Pavement. The bricks for this were made of washed and prepared clay and sand, and after being baked they were rubbed to the proper size and shape. There is a very fine old brick house at North Wheatley, dated 1673. In the Middle Ages the forest of Sherwood supplied an abundance of timber for the frame houses.

Plaster made from gypsum was used from Roman to recent times for making floors, and many examples of these still exist.

Over the greater part of the county, owing to the abundance of clay in the coal-measures, Permian, and Keuper formations, bricks and tiles are the commonest house-building material. The majority of village churches however are built of local stone. In the clay country this is from either the Skerry bands or the Waterstones. The

Governor's Old House, Newark

latter was used for old Nottingham Castle and outer portions of Newark Castle; whilst the inner portions were built of Lias limestone. None of these stones are now used to any extent.

In the west the Magnesian Limestone is used for private as well as public buildings. Because it had lasted so well in Southwell Cathedral this stone was used for the Houses of Parliament.

There has been little need to import stone from other counties. Oolite was brought from Lincolnshire for the older parts of Newark Castle, and for Wollaton Hall and University College. Much slate for roofing has been introduced, but happily it has not succeeded in completely displacing tiles even in the newest houses.

21. Communications—Past and Present.

Nottinghamshire was crossed in Roman times by several roads. Of these the Fosseway, made about A.D. 120, is the best preserved. It crossed the country from Bath and passed through Leicester to Lincoln. It entered this county with the wolds and left it north-east of Newark. It kept close to the crest of the highest ground south of the Trent and consequently had a dry foundation. Another road, Till Bridge Lane, branched off from the Ermine Street just north of Lincoln, crossed the Trent at Littleborough by a pavement which still exists, and pursued a north-westerly course to Bawtry. Leeming Lane, north of Mansfield, is the remnant of another road, which entered the county near Sutton-in-Ashfield and also left it at Bawtry.

Next in age to the Roman roads came the North Road. From Rempstone it makes its way north through Nottingham, Ollerton, and Blyth to Bawtry, and thence to York. At Nottingham, where it gave names to Stoney Street, Broad Street, York Street, it has become part of Mansfield Road. It was along this road that the Danes came in A.D. 868. By the thirteenth century it had ceased to be a noted road. Another one had come into existence, parallel to it, but passing through a set of more important towns. This went from Nottingham through Mansfield with its royal manor Warsop, and Worksop with its great priory.

The Great North Road enters the county just south of Balderton, crosses the Fosseway at Newark, passes through the old post-town of Tuxford, and on to Retford and Bawtry. Its name is more familiar than that of the North Road because it became the main trunk to the north during the "coaching days." This was partly due to the fact that a greater number of places lay along its route. Between Newark and Bawtry it passes eighteen, whereas the other road in a longer distance, viz. from Nottingham to Bawtry, passes only four. The posts for Nottingham were delivered from Newark.

Thus four ancient roads converge from Littleborough, Sutton-in-Ashfield, Rempstone, and Balderton upon one point, Bawtry. It has been already stated that all the country between this town and the Trent and Humber was covered with swamps. Bawtry was the most easterly

The Great North Road, leaving Newark

point at which these roads could cross the Idle and pass round the swamps on their way to the north. Where they enter the county they are widely separated. This is due to the great distance between the gateways into the plain of Nottingham at Lincoln, Grantham, and Loughborough.

The winding courses of the North and Great North

Sutton on Trent, Great North Road

Roads stand out in marked contrast to the straightness of the Fosseway. The latter was made for military purposes. The two former were probably made by piecing together pre-existing roads and tracks.

From the introduction of railways until the invention of motor cars the necessity for great trunk roads was not felt. But at one time every town and village was a centre from which good roads radiated in all directions. No doubt many of these existed a century and a half ago, but in those days they were so badly kept that goods had to be carried about on pack-horses and a speed of three miles an hour was considered satisfactory.

Towards the close of the eighteenth century the in-adequacy of the roads for the purposes of transport was partly made up for by the making and improvement of waterways. Nottinghamshire was favoured in the possession of the Trent. It seems very probable that this river was used by the Romans at least from East Bridgford downwards. In the fourteenth century it was so important a means of transport that Nottingham became a noted river port. In the eighteenth and at the commencement of the nineteenth century goods from all parts of the country were conveyed by it to the Humber for south Yorkshire or London, and to the Fossedyke for Boston, London, and France.

After the introduction of railways the traffic on the Trent declined, but during the last twenty years attention has been turned to it once more. The channel has been deepened by dredging and the making of weirs, and steam or motor traction has been introduced. These improvements have already enabled it to become an efficient competitor with the railways. Vessels carrying as much as 150 tons come up to Newark and, when recently-granted powers to construct additional locks and weirs have been used, such vessels will be able to come fully loaded up to Nottingham also. Thus the river is gradually regaining its old position as an important commercial highway. Should the great "four rivers scheme" for connecting the Trent, the Severn, the Thames, and the Mersey in the neighbourhood of Birmingham be carried through, its importance will be considerably enhanced.

In its northerly reaches the Trent is connected with the flourishing network of canals in the West Riding of Yorkshire. Within the county it receives the Fossedyke,

Colwick Weir

the Chesterfield, the Nottingham, and the Grantham canals, all of which are owned by the railways and should be commercial feeders of the Trent. Above Nottingham it is connected with London through the canalised river Soar and the Grand Junction Canal, and with the Black Country and Liverpool by the Trent and Mersey Canal.

The waterways helped forward the commercial development of England considerably during the period 1750 to 1850. The railways have done the same, but to a much greater extent, during the last sixty years. The county possesses an intricate network of lines belonging mainly to the Midland, Great Northern, and Great Central Railways. The first line, which was made by the present Midland Company from Nottingham to Derby, was opened in 1839. Several influences have governed the successive extensions which have since been made. Foremost among these has been the presence of a great coalfield in the west. During the earlier years of railway enterprise the Midland and Great Northern were anxious to secure access to this field and through communication to the London market. That these objects have been attained is shown by the numerous lines entering the county from the west and uniting to form main lines to the south.

Three lines strike across the county to the east and converge on the Lincoln gap. These supply northern Lincolnshire and the port of Grimsby.

The presence of great stores of ironstone south of the Belvoir Escarpment influenced the making of two lines from Nottingham to Melton Mowbray.

Underlying all these influences was the desire to make through routes from London to the North and Scotland. Two of these belong to the Midland Railway. The one skirts the county along the valleys of the Soar and Erewash, the other comes through Melton and Nottingham and passes north along the Leen Valley. The Great Central main line leaves the Soar Valley near Loughborough and

The Trent, near Cromwell

strikes across to Nottingham and the north along the same valley. The Great Northern main line runs along the same route as the Great North Road and is connected with Nottingham by an important branch which runs into Derbyshire and Staffordshire.

The London and North Western and the Great Eastern have running powers in some parts of the county.

22. Administration—Past and Present.

The early Teutonic settlers in our county lived in towns or, as they would now be called, villages. A town was an enclosure surrounded by a hedge or a palisade. The inhabitants co-operated with one another in the cultivation and use of the neighbouring land. The whole of such an area was a township. The agricultural affairs of the township were discussed at meetings presided over by the town-reeve or the lord's steward.

For the detection and punishment of crime, and to facilitate the levying of taxation, the townships were grouped into Hundreds or Wapentakes. The latter were a Danish institution and received a Danish name. In the region of this county they were so arranged that the Trent formed the boundary between two sets, whilst the Fosseway and North Road passed through them. Each had its own council or *moot*, consisting, in theory, of the lord, the priest, and four representatives from each township. This moot was usually held on or near a prominent hill.

About the tenth century our shire was formed. Until the twelfth century it contained eight wapentakes, viz. Rushcliff, Bingham, Newark, Broxtow, Basset-law, Oswardbeck, Lida, and Thurgarton. Subsequently Oswardbeck was combined with Bassetlaw, and Lida with Thurgarton. The shire was ruled by an Earl appointed by the Witan or moot of the kingdom. The Sheriff, an officer of great importance, was chosen by the King.

In these divisions and their respective moots were the germs of the present system of government. Then, as now, there was a combination of local with central government. The functions of this have now become more clearly defined and may be classified under three heads, viz. poor-relief, general welfare, and justice.

With the introduction of Christianity came the priest, who naturally took up his abode in the town and looked after the spiritual welfare of one or more townships. The area of his activities became his parish. To the parish meeting all the business originally done at the township meeting in course of time gravitated. During the last century acts were passed which separated secular from ecclesiastical affairs and established civil parishes. These generally correspond to the original townships. In Not-tinghamshire there are nearly 250 civil parishes. When the population is less than three hundred the business of the parish is done at a meeting of the ratepayers called the parish meeting. If the number is greater it is done by an elected body, the parish council.

To economise the cost of carrying out poor-relief several civil parishes, whilst having their own poor-rate, may unite and form a Poor-Law Union. Of these there are eight, viz. Basford, Bingham, East Retford, Mansfield, Newark, Nottingham, Southwell, and Worksop. The work of the union is done by the Board of Guardians, who consist of at least one guardian elected from each parish.

Matters which affect the general welfare of the community, e.g. roads, drainage, education, etc., are dealt with by the District and County Councils. Each district consists of one or several civil parishes. In some the population is sufficiently great to impart the characteristics of a town (in the modern sense). These are called Urban Districts, e.g. Worksop and Hucknall. The others are called Rural Districts and have the same boundaries as the Poor Law Unions except where these overlap the Urban Districts.

Nottinghamshire became an administrative county in 1889. Its council consists of a chairman, 17 aldermen, and three times that number of councillors.

Two other important areas have yet to be mentioned, viz. Municipal Boroughs and County Boroughs.

The former is an urban district which has received a charter from the Crown. It has a council of its own made up of mayor, aldermen, and councillors. It has other functions in addition to those of an ordinary district council, e.g. it may have its own police force. The Municipal Boroughs of Newark and East Retford are very ancient; that of Mansfield is modern.

The City of Nottingham is one of the sixty-four County Boroughs in the country. As such it is exempted from the control of the County Council and has all the privileges and powers of a county itself. Its council has the same constitution as that of a Municipal Borough.

For Parliamentary elections Nottinghamshire is divided into four divisions, each of which sends one representative. The Parliamentary Borough of Nottingham happens to

The Guildhall, Nottingham

have the same boundaries as the County Borough. It has three divisions and three representatives.

All the administrative bodies so far considered are elected by the people of the locality. Those who administer justice are not so elected but are appointed by the King. In the administration of justice cases brought by one person against another are distinguished from those brought by the Crown against breakers of the law. The former are civil, the latter criminal cases.

For dealing with criminal cases about one hundred and ninety-three justices of the peace have been appointed in Nottinghamshire. Only two justices are necessary to hold the Petty Sessions for the punishment of minor offenders. One justice may decide whether there is ground for more serious accusations to be dealt with at the Assizes or Quarter Sessions. The latter are held four times a year at Nottingham, Newark, and Retford, and must be attended by a number of justices.

The more important civil and criminal cases are tried before judges of the Supreme Court at the Assizes. These are held at various towns which are grouped according to the density of the population into circuits. Nottinghamshire is in the midland circuit. The County Court is presided over by a judge and deals only with minor civil cases. For County Court purposes England is also divided into special circuits. Nottinghamshire and a portion of the West Riding of Yorkshire form County Court Circuit No. 18. In this circuit there are eight districts, of which six are in this county, viz. Bingham, Retford, Mansfield, Newark, Nottingham, and Worksop.

Ecclesiastically Nottinghamshire is now in the province of the Archbishop of Canterbury, though formerly it was in that of York. In 1836 it was transferred from the diocese of York to that of Lincoln. In 1884 it was joined with Derbyshire, which belonged to the diocese of Lichfield, to form the diocese of Southwell.

In early times a group of villages in the north of the

shire was exempted from the local organisation of the county and the jurisdiction of the Sheriff. This constituted the Liberty of Southwell and Scrooby. Within this group exclusive privileges in all legal matters were granted by the Crown to the Archbishop of York. The Court was held at Southwell.

23. Roll of Honour.

Situated as it is on the direct and easiest route from the south to Scotland, and possessed of many attractions of its own, Nottinghamshire was often visited by the early kings, especially by John and Richard III, with whom the castle was a favourite residence. It was from here that the latter went forth to Bosworth Field. The former died in the companion castle at Newark. Edward I frequently visited the county on his way to and from his wars in Scotland. It was during one of these visits that his beloved Queen Eleanor died at Harby.

Southwell with its cathedral was situated at the southernmost point in the province of the Archbishop of York. Consequently the great ecclesiastics often occupied the beautiful palace here. The most noted amongst them was Cardinal Wolsey, who retired to this place after his downfall.

The county has produced several archbishops. Chief amongst them was Archbishop Cranmer, who was born and spent his boyhood at Aslockton. The first bishop of the newly-created (1884) bishopric of Southwell was Dr Ridding.

William Brewster, the leader of that Puritan band who ultimately sailed in the *Mayflower*, was for a long time a resident and postmaster at Scrooby.

The village of Cromwell gave its name to a famous

Archbishop Cranmer

family. One of them, Ralph, Lord Treasurer under Henry VI, was intimately associated with the county. This was not the case with Thomas Cromwell, the minister of Henry VIII, and Oliver Cromwell, both of whom were distinguished members of that family.

Of prominent Parliamentarians, Denzil, Lord Holles, was born at Haughton near to Retford. It was he who on one occasion held the Speaker in his chair whilst the House passed resolutions against the unjust axes and religious innovations of Charles I. He was also one of the five whom the King accused of treason. John Evelyn Denison, Speaker of the House of Commons (1857–72), was born at Ossington.

Colonel Hutchinson, who held Nottingham Castle during the Civil War, and General Ireton, Cromwell's son-in-law, were both well-known names in the county. The former came from Owthorpe and the latter was born at Attenborough.

With the village of Langar are associated two great families—the Scropes and the Howes. A stately memorial to Admiral Earl Howe, who distinguished himself in the wars with France, stands in St Paul's Cathedral, but his remains lie peacefully in the village church of his ancestral home.

Several famous Elizabethan sailors hailed from this county, notable among them Sir Hugh Willoughby, who attempted to find a north-east passage to China and perished with all his companions in the attempt. He sprang from a distinguished family at Willoughby on the Wolds, which now for many generations has resided at Wollaton Hall. Sir Martin Frobisher, explorer of the north-west passage, belonged strictly to Yorkshire, but he owned the manor of Finningley. He was accompanied on several occasions by Edward Fenton, who came from the hamlet of Fenton, near Sturton.

The men of Nottinghamshire have also served their country well by their inventions. By inventing the stocking frame the Rev. William Lee of Calverton became the father of the hosiery industry. John Heathcoat invented a machine for making bobbin net, and thus saved the lace industry for Nottingham. The great cotton

St Mary's, Hucknall Torkard
(*Byron's burial place*)

industry is mainly dependent upon two machines, the spinning-frame and the power-loom. The former was invented by Sir Richard Arkwright, whose first mill, driven by horses, was set up in Hockley, in Nottingham. The latter was invented in 1784 by Edmund Cartwright, a member of a well-known family which resided at Marnham. Mechanical ingenuity was not his only gift for he was a divine of no mean order, a great scholar, and somewhat of a poet. Of the same spacious type of mind was Dr Erasmus Darwin of Elston Hall, "poet, physician, philosopher, philanthropist," and grandfather of

Memorial to Kirke White in Wilford Church

Charles Darwin. He also attempted to devise various mechanical contrivances, for example "an organ which should pronounce the Lord's Prayer, the Creed, and the Ten Commandments."

Marshall Hall, a physician renowned for his researches

on the nervous system, was born in Basford and practised in Nottingham.

For centuries the name Byron shone in connection with this county. It died out in sunset splendour with the poet. Several years of his youth were spent at South-

Richard Parkes Bonington

well. For the next few Newstead Abbey was his home. He died at Missolonghi in Greece in 1824, and his body was brought to England and interred in the family vault in Hucknall Torkard church. Of lesser poets Henry Kirke White was born the son of a Nottingham butcher and died at twenty-one, and Philip James Bailey, the author of *Festus*, was a native of Nottingham. The well-known writers William and Mary Howitt, though not natives of the county, lived many years of their busy literary life in Nottingham. The county historian, Thoroton, belonged to the village of Car Colston.

In the domain of Art there is one name which stands out pre-eminent. Richard Parkes Bonington, who was born October 25, 1802 at Arnold, son of the Governor of Nottingham prison, had but a short life, for he died in 1828, but he attained the very highest excellence both as a water-colourist and as a painter in oils, and year by year the appreciation of his genius gathers strength. At Nottingham, in 1725, was born Paul Sandby, the father of English water-colour painting, and four years previously his less-known brother Thomas.

24. Towns and Villages. Their Names and Distribution.

Our Teutonic ancestors were not merely warriors. They fought, not for the sake of fighting, but to secure for their families homes to live in and lands to cultivate. From them the great majority of our towns and villages received their names. Such endings as *ton*, *worth*, *ham*, *borough*, *stoke*, *stowe* take the imagination back fifteen hundred years, and conjure up the picture of an Anglian group of settlers marking off a piece of ground for its own and protecting it from sudden attack by planting a hedge or erecting a stockade.

To this enclosure the name of the chief man of the settlement was often given, e.g. Beckingham and Ruddington, from the personal names Becca and Rudda respectively. The endings *ley* and *field* raise pictures of peaceful occupations. The former signifies a pasture or forest glade, the latter a clearing made by felling trees. Sometimes a name contains a record of former features of the countryside. Thus *ey* in Mattersey depicts this place as an island in the midst of a sea of swamps, but *holme* was the Scandinavian for an island enclosed by a river. Round Sutton-in-Ashfield ash trees abounded. At Farnsfield ferns were numerous. Beavers built their little mud huts at Bevercotes. Halam means "at the meadows"; Harwell, "hare spring"; Shelford, "sloping ford"; Worksop, "Weorc's valley."

As communities grew and extended "North," "South," "East," and "West" were prefixed to some place-names. In some cases these became altered almost beyond recognition, as in Sutton and Norton. The sites for enclosure were chosen for the sake of certain natural advantages they offered—a good supply of water, proximity to fertile soil, freedom from floods, etc. It is not difficult to discover the causes which produced these advantages and therefore governed the choice of sites.

In the accompanying map the shape, size, and position of every town and village in Nottinghamshire is shown. A glance at it reveals at once their scarcity in the Bunter country. Here the soil was not sufficiently fertile nor the water accessible enough to offer any great attractions to settlers. The few villages that are found are situated

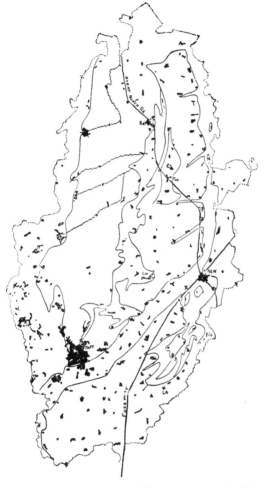

Map showing distribution of Notts. towns and villages

(*The fine lines mark off boundaries of geological formations as far as
these have influenced the distribution. The wavy lines are streams
on Bunter outcrop*)

either on the rivers or on the extreme eastern margin of
the Bunter, where water is sometimes ten feet below the
surface.   Thus it came about that in this area much
primeval forest and waste were left unenclosed and became
a royal hunting ground.   On the other hand its sylvan
beauty and quietude attracted the monks, who established
many of their monasteries either within or close to its
precincts.   In later times extensive portions passed from
the hands of king and clergy and became those great
landed estates now known collectively as the Dukeries.

The outcrop of the Waterstones presents a marked
contrast to that of the Bunter.   Consisting as it does of
layers of sandstone and clay, it yields a fertile loamy soil
and an abundant supply of water which can be obtained
from springs or by means of shallow wells.   These two
advantages attracted the early settlers and account for the
many villages between Retford (R), Southwell (S), and
Nottingham (No).   Occasionally a village, like Lambley,
stands mainly on the Keuper clay but gets its water from
wells sunk through this into the Waterstones.

The outcrop of the Keuper bears many streams, but
portions of these are liable to failure during spells of dry
weather.   The sandstones of the Skerry bands yield a
better and more constant water supply, and also produce
a lighter soil than does the clay.   The presence of an
extensive area of this stone has furnished sites for a group
of large villages north and south of Tuxford.   In the Vale
of Belvoir many of the villages are situated on that margin
of the Keuper which lies next to the alluvium and stands
sufficiently above this to lift the villages beyond the reach

of the floods of the Smite. The string of villages from Langar (L) to Cotham (C) stands upon the outcrop of the Lias limestone. This is more resistant than the Keuper and Lias clays, and consequently has formed a range of very low-lying hills. Here, then, the villages have a suitable foundation, a supply of comparatively good water drawn from the limestone, and a situation elevated above the flood levels of the river.

Within the vale of the Trent also the sites of the villages have been largely determined by the necessity for escaping from the floods. With few exceptions they all stand upon those patches of gravel which, as shown in an earlier chapter, rise ten or twenty feet above the surface of the alluvium (p. 145). This elevation is sufficient to save the villages from all but the severest floods. At the same time the gravel provides a dry foundation and does away with the necessity for wasting the fertile alluvium in building sites.

Up to 1850 the southern building line of Nottingham coincided with the boundary between the Bunter and the alluvium. At that time the latter was used as a pasture by the citizens. Even now that it has become valuable for building purposes it is still called " The Meadows."

In t e Coal-measure and Permian areas the villages are numerous, but have lost their original compact character. Buildings have crept out along the roads, whilst the intervening fields have been retained for allotments and agriculture. Thus typical straggling colliery villages have grown.

In the early days each village was self-contained and

its inhabitants had little to do with those of the neighbouring villages. Whilst therefore the Fosseway provided a good means for penetrating the country, settlements were made several miles from it and at the centre of the area to be cultivated. The road was, and still is, used as the boundary between townships and parishes. A thousand

Scrooby from the North. Great North Road on Right

years later all the villages were knit together by a sense of common nationality and by many common interests. The main highway at this time was called the Great North Road because it passed through so many towns and villages. The North Road, though more ancient, fell, as we have seen, into comparative disuse for the contrary reason.

25. The Making of Nottingham.

After considering the geography of the county it will
not be out of place to summarise the influences which
have controlled the origin and growth of that city which
gave the county its name and now contains nearly as
great a population as all the other towns and villages
combined.

In the last chapter it was seen that sites must have
been chosen because they possessed certain natural ad-

Nottingham in the middle of the Eighteenth Century

vantages. This is equally true of Nottingham. The
accompanying picture shows the city as it appeared about
the middle of the eighteenth century. Then and in
previous centuries it was situated wholly within a half-
saucer-shaped hollow, with the castle rock like a bastion
at one end and St Mary's hill at the other. Between was
a semicircle of high ground which sheltered the hollow
from the cold north and north-easterly winds and gave it
a southerly aspect. Here, then, was a warm and sheltered
site with places on either side which could be easily
fortified and used as refuges.

On the south and west ran the Trent and Leen with their unfailing water-supply and abundance of fish. Away to the north and east stretched the forest, whence could be obtained wood for fuel and game for food. The underlying sandstone was soft enough to be easily excavated into caves and dry enough for these to be habitable. Though the soil was indifferent the situation

Nottingham Castle in the Sixteenth Century

was unique, for within a radius of three miles was to be found, with one exception, every type of rock and consequently of soil within the county.

Whilst the city offered these advantages to the inhabitants it occupied a position of strategic importance at a time when the country was divided into little kingdoms struggling one with another for supremacy. At that

time the Trent was a formidable barrier to the move-
ments of armies. If an army came from the north it
naturally avoided the hill-country of the Pennines on the
west and the swamps on the east. Beyond the river its
path was barred by the precipitous nature of the opposite
side of the vale and by the Belvoir Escarpment.   Only
one of two courses remained,—to cross the Trent at

Trent Bridge, Nottingham

Newark and go up the valley of the Witham, or to cross
it at Nottingham and use the valleys of the Soar or Upper
Trent. As the powerful kingdoms of Mercia and Wessex
lay to the west and south the latter course would be the
more usual.

It is not surprising that to these natural advantages
artificial ones were soon added. By the ninth century a
great road to the north had come into existence. This

kept strictly to the dry sandstone country and was thus
more passable at all seasons than if it had been made upon
the clay. It struck this country at its most southerly limit,
namely at Nottingham. By the eleventh century the
importance of the situation was still further enhanced by
the addition of a bridge and a Norman fortress. For a
long time Nottingham's central position in the kingdom,
its magnificent castle, and the abundance of sport in the
neighbouring royal forest caused it to be the frequent
resort of the kings. They were usually accompanied by
a large retinue. Many people would also congregate here
in consequence. Such a place was a suitable one for
holding markets and fairs, and provided opportunities to
exchange merchandise at a time when shops were non-
existent. It thus became a trading centre.

Its position near to a navigable river gave it more
than local importance. Throughout mediaeval and early
modern times carriage overland was difficult and expensive,
especially for bulky goods. During this period Nottingham served largely as a river port for the Midlands.
Such things as wool, pottery, lead, coal, and gypsum were
carried down, and machinery, etc., brought up the river.
The making of canals and the canalising of smaller rivers
at the end of the eighteenth century only enhanced its
importance by furnishing more efficient feeders to its
waterway.

Meanwhile industries grew up and flourished. In the
development of these the neighbouring forest played its
part, for it supplied oak-bark to the tanners, excellent
close-grained wood to the carpenters and cabinet-makers,

and fuel to the smelters of the iron ore from the Erewash valley. These last did not have their forges in Nottingham, but they supplied its smiths and armourers with iron. "The little smith of Nottingham" was far famed. It is probable that he derived his skill in metal work from the Danes.

That William Lee, the inventor of the stocking frame, lived only a few miles away was an important circumstance. Under his care a band of trained workers came into existence. Though he went to seek his fortunes elsewhere, these remained and nurtured the young industry in this district. Gradually it spread from village to village and also to Nottingham. Here many of the workers were not content merely to make stockings but, with the advantages of inherited mechanical skill, they strove to improve the machine and adapt it to other classes of work, more especially to the making of lace. Thus improvements too numerous to mention were added from time to time, often by humble and obscure workmen. Their efforts laid the foundations of an industry which has made Nottingham the lace metropolis of the world.

When hand power was replaced by water power these and other minor industries might soon have left the city, for the Leen was too small to drive many mills. However, the reign of water power was short. Steam power was introduced. This depended entirely on a good coal supply, and Nottingham had a coal field at its door and coal in its foundations. With steam power came the large factory and the consequent rapid concentration of the population in manufacturing centres.

The importance of the place and the easy natural gradients which lead to it from all directions soon attracted the railways. The network of main and branch lines which now serve it will never be easily shifted and consequently adds greatly to its commercial stability.

The gradual eastward extension of the collieries and attendant industries is enveloping the city on the north with a densely populated district. This is adding to the city's prosperity, for even now people come from the Erewash valley and from beyond Mansfield to make their purchases and seek their pleasure in Nottingham.

26. THE CHIEF TOWNS AND VILLAGES OF NOTTINGHAMSHIRE.

(The figures in brackets after each name give the population in
1901 and those at the end of each section are references to
the pages in the text.)

Annesley (1271), six and a half miles S.W. of Mansfield,
on the borders of Sherwood Forest. Annesley Hall is a fine
example of a country mansion. It was the birthplace and home
of Mary Chaworth, who was immortalised by the poet Byron. It
is now the seat of the Musters family.

Arnold (8757), four miles N.E. of Nottingham, is a busy
centre for the manufacture of lace and hosiery. Brewing
and soap-making are also carried on. It is the birthplace of
Richard Parkes Bonington, one of England's greatest painters.
The Early English portion of the fine parish church was built in
1270. Cockpit Hill, on which are the remains of a Roman encampment, is close by. (pp. 57, 67, 121.)

Aslockton (372), on the Smite, is three miles east of
Bingham. Archbishop Cranmer was born here in 1489, and here
he received his early education in learning and manly sports.
"Cranmer's Mound" is really a part of some ancient earthworks and consists of two rectangular courts surrounded by a foss.
The latter is now nearly filled up and the shape of the mound
spoilt by the removal of material for ballast. (pp. 79, 80, 115.)

Beaumond Cross, Newark

Attenborough (1176), lying five miles S.W. of Nottingham, was the birthplace of General Henry Ireton (1611) son-in-law of Oliver Cromwell and Lord Deputy of Ireland. His brother John became Lord Mayor of London 1658. The population includes Chilwell. (p. 117.)

Balderton (2203), near the Witham, and two miles S.E. of Newark, has some large engineering works. The church contains some good examples of Norman work, especially in the porch. (pp. 82, 103.)

Basford (27,119) is an important centre for the hosiery industry. It was incorporated in the borough of Nottingham in 1877. It was the residence of Philip James Bailey and the birthplace of Marshall Hall, and is head of a Union. (pp. 112, 121.)

Beauvale, in Greasley parish, seven miles N.W. of Nottingham, is the site of a Carthusian monastery which was founded here in the reign of Edward III. The site is now occupied by a farmstead where portions of the old buildings may still be seen. (p. 90.)

Beeston (8960), lies near the Trent three miles S.W. of Nottingham. The population increased from 6948 between the last two censuses. The older parts were situated upon a terrace of ancient river gravels and thus stood above the flood level. Owing to the improvement of the river's channel and to better drainage the great expanse of alluvium between it and the Trent is being rapidly covered with buildings and factories. The Midland Railway has sidings for the coal brought down from the Leen Valley Collieries. Other industries are lace and telephone making, malting, and foundry work. (p. 57.)

Bestwood Park (650), once part of Sherwood Forest, belongs to the Duke of St Albans. Bestwood Lodge was built about 1858 in the fifteenth century style. In the district there are a modern colliery, blast furnaces, and one of the Nottingham Corporation pumping stations.

Bingham (1604). A market town near the Fosseway, ten miles east of Nottingham, which depended for some of its prosperity upon the fact that it is situated near the centre of the fertile corn-growing Vale of Belvoir. It gave its name to one of the ancient wapentakes. Moot House Pit was probably the place of meeting. The Rt Hon. Robert Lowe, Chancellor of the Exchequer and afterwards Viscount Sherbrooke, was born here, his father being rector of the parish. It is a Union town, head of a Petty Sessional division and County Court district. (pp. 89, 111, 112, 114.)

Blyth (571), six miles N.E. of Worksop, on the Ryton. It was one of the five places in the kingdom licensed to hold tournaments. A Benedictine monastery was founded here in 1088. During the Middle Ages this was an important place for the entertainment of travellers along the North Road. For this reason the road was probably diverted from its more direct course so that it might pass through Blyth. The monastery and the choir and tower of the church were long ago destroyed. The portions which now remain are some of the earliest Norman work in the county. Blyth was formerly an important market town but the market was removed some years ago to Bawtry. (pp. 80, 86, 90, 103.)

East Bridgford (756), a village on the Trent, nine miles N.E. of Nottingham, close to which is the site of the Roman station Margidunum, which was situated on the Fosseway. Here the road comes close to the Trent for the first time on its way from the south-west. From it a short road, now known as Bridgford Street, runs down to the river. Goods were probably shipped here and sent down the river to York or Boston.

A beautiful kind of gypsum, called satin spar because of its lustre, was at one time found here in layers six inches thick, and was used for making ornaments, beads, etc. (pp. 65, 77, 106.)

West Bridgford (7018). A parish two miles south of Nottingham which shows considerable rate of increase of population during the nineteenth century. Moreover, for the period 1897–1906 the average gross death-rate for the district was only 8·29 per thousand, which is much lower than for any other district in Nottinghamshire.

Brough. A tiny hamlet on the Fosseway, one mile east of Langford. Here, perhaps, was situated the Roman station Crococolanum mentioned by Antoninus in his *Itineraries*. Numerous coins, fragments of pottery and implements, and foundations of houses have been found over an area of forty acres. Probably it was not merely a resting stage for soldiers but a settlement of some importance. (pp. 77, 99.)

Bulwell (14,767) was incorporated in the borough of Nottingham in 1877. Bulwell Forest is a remnant of the old forest waste which is now used as a public recreation ground. The usual manufacturing industries of the district are carried on here. The stone from extensive quarries in the Magnesian Limestone is much used for making burnt lime and for building walls. Equally extensive pits in the Permian marl supply clay for bricks, tiles, and flower-pots. Moulding sand is obtained in large quantities from the Bunter. The localisation of these industries here is largely due to the proximity of the railways. There are also coal-pits. (p. 58.)

Calverton (1159) in South Notts, six and a half miles N.E. of Nottingham, is claimed to be the birthplace of William Lee, inventor of the stocking machine. Hosiery is still made here, but not in such large quantities as formerly. Traces of Saxon or Early Norman work exist in the church. In Foxwood, half-a-mile south of the village, are some earthworks which formed a stronghold in easy communication with a number of minor encampments in the surrounding district. (pp. 57, 67, 118.)

Clipstone, on the Maun, three miles N.E. of Mansfield, is the site of a castle often used by royal hunting parties during the twelfth and thirteenth centuries. It was so frequently occupied by King John that even now the ruins bear the name "King John's Palace." (p. 95.)

Clumber. The seat of the Duke of Newcastle, three miles S.E. of Worksop, with a park 11 miles in circumference. The mansion was erected about 1772, and in 1879 it was partially destroyed by fire. (p. 99.)

Eastwood (4815), nine miles N.W. of Nottingham. It was here that the colliery owners met in 1832 to consult about the loss of the Leicester market for the sale of their coal, which was due to a railway made to that town from the Leicestershire coalfield. It was thereupon decided to lay down a line also from the Erewash Valley coalfield. Though it was many years before the line was made this meeting was the germ whence the great Midland Railway ultimately grew.

This parish showed the greatest percentage increase, viz. 63·50 per cent., in population per square mile in this county during the nineteenth century. This was associated with the growth of the collieries. Eastwood is situated in the rainiest part of the county on its extreme western border. (p. 62.)

Edwinstowe (904), a township seven miles N.E. of Mansfield, named after Edwin, King of Northumbria, who was slain in the battle on the Idle, A.D. 633. It is a favourite centre from which to visit the Dukeries.

Gotham (1009), lying seven miles S.W. of Nottingham, is a name made familiar to many by the stories of "The Wise Men of Gotham." It is a centre for gypsum mining and plaster making. (p. 28.)

Gringley on the Hill (720) lies six miles E.S.E. of Bawtry. The top of Beacon Hill is the finest view-point in the county. The outlook embraces Lincoln Cliff, the Carr lands as far as the

Ouse, and the country west of the Magnesian Limestone escarpment. Here are earthworks believed to date back to Roman or British times. Prince Rupert encamped here before going to the relief of Newark in 1644. (p. 10.)

Hucknall Torkard (15,250). A prosperous town on the Leen seven miles N.W. of Nottingham. Many of the men are engaged in the mines on the neighbouring estates of Annesley, Newstead, and Bestwood. Laundry work, cigar-making, and hosiery-manufacturing are carried on here. The making of hosiery shawls is an old industry. Lord Byron was buried in the chancel of the parish church. (pp. 57, 60, 112.)

Kimberley (5129), five miles north of Nottingham, is a busy place, with brewing, mining, and knitting industries. The railway cuttings here are very instructive as they show the Permian rocks resting on the coal measures.

Kingston-upon-Soar (271), ten miles S.W. of Nottingham, is the centre of an important dairying district. In 1900 the agricultural department of University College, Nottingham, was removed here and formed with the existing Dairy College "The Midland Agricultural and Dairy College." Various general and special courses are arranged to meet the requirements of different types of students. In addition much is done by consultations, analyses, and a travelling Dairy School to advance these industries in the co-operating counties. (pp. 53, 79.)

Laxton (or **Lexington**) (394), is a village lying three and a half miles S.S.W. of Tuxford. Half-a-mile to the north are the largest and best preserved earthworks in the county. They consist of a mount and two courts and baileys. The mount is capped with a tumulus and surrounded by a foss. The courts are also bounded by a foss. (p. 22.)

Lenton (14,662), was included in the borough of Nottingham in 1877. Formerly a great priory of the Cluniacs existed here, but only traces of the building remain. A beautiful Norman font is still preserved in the parish church.

Littleborough (49), now an insignificant place, is situated six miles east of East Retford, at the point where the Roman Road, Till Bridge Lane, crossed the Trent. Remains of the pavement still exist in the floor of the river, and other Roman relics have been found. This is probably the Segelocum or

The Southern Boundary of Old Sherwood Forest

Agelocum of the *Itineraries* of Antoninus. The church here is Norman in style with Early English additions, and is one of the smallest in the county. (pp. 79, 102.)

Mansfield (21,445) is named after the river Maun upon which it stands. It was probably a Roman station situated at the point where the old road which ran north of Southwell crossed Leeming Lane. During the Middle Ages it was often the resort of royal hunting parties. At a later time quarries of excellent building stone were opened in the neighbourhood. To the east lay extensive areas of corn- and barley-growing country. On the west was the Erewash Valley coalfield. These influences helped to make it an important market town at an early date. Before railways were introduced a horse-tram ran to Pinxton and was connected with the Cromford and the Erewash Valley canals. These brought coal and cotton and carried back stone, lime, and corn. It is now almost surrounded by collieries, and is a busy manufacturing centre for hosiery and boots, lace-thread spinning, and cigar-making. It is the head of a Poor Law Union, and a Petty Sessional Division. In 1891 it became a municipal borough. (pp. 46, 57, 60, 66, 102, 112, 114.)

Mansfield Woodhouse (4877) is one and a half miles north of Mansfield. Remains of a Roman villa have been found here. In the immediate neighbourhood are numerous stone quarries. (pp. 77, 89.)

Misterton (1433), a village on the Chesterfield canal five miles N.W. of Gainsborough. There are chemical and copper-precipitating works here. Close by is Misterton Soss, where the pumping station for the Morther drain is situated. (pp. 12, 35, 36.)

Netherfield is three miles E. of Nottingham. Here the Great Northern Railway has very extensive sidings where iron-stone from Leicestershire and coal from the York, Derby, and Nottingham coalfields are dealt with.

Newark (14,992) after its destruction by the Danes about 1041 was rebuilt, hence the name New werke. It stands on the Devon, which here has the appearance of being a loop of the Trent, and is situated at the point where the natural route from London down the valley of the Witham to the north crosses the Fosseway and the Trent. Here Bishop Alexander built a castle in the twelfth century which was further strengthened in the

thirteenth. These factors helped to make it a position of great
strategic importance during the civil wars, when it resisted several
vigorous sieges. In the coaching days its position on the Great
North Road led to its becoming a famous post-town. The coming
of the main line of the Great Northern Railway and the Notting-
ham-to-Lincoln branch of the Midland Railway added considerably
to its advantages, while the Trent is also a great aid to its trade.
It possesses works for the manufacture of machinery, agricultural

Old Grammar School, Newark

implements, artificial manures, and flour, while malting and
brewing are carried on extensively. Much gypsum is mined and
limestone quarried in the district. As a market town it stands
second only to Nottingham. The castle ruins, parish church, and
a number of old half-timbered houses make it well worth a visit.
It is a municipal borough, incorporated under Edward VI, and has
had a mayor since 1625. It is head of a Poor Law Union and
Petty Sessional Division. (pp. 61, 102, 111, 112, 114, 115.)

Newstead (1100) is five miles south of Mansfield. Newstead Abbey was founded by Henry II, it is said, as an act of
penance for the murder of Becket, and was occupied by the
Austin Canons. At the Dissolution it passed to the Byron family
and thus became associated with the poet. Under the last two
representatives of that famous line it fell into almost hopeless
disrepair. In 1818 it was purchased by Colonel Wildman, who
carefully preserved what remained of the old part and added
enough to make it a charming residence. (pp. 12, 88, 90, 97, 120.)

Nottingham (239,743). Situated on the Trent, and the
capital of the county. Entered by the Danes in 868 it became
one of the Five Danish Boroughs. Its castle was built by
William I in 1068. It received charters from various kings
beginning with John, and became a county in itself in 1448. As
a Parliamentary borough it returns three members. It is a great
manufacturing and distributing centre, especially noted for lace,
and also to a less extent for hosiery. Many other industries are
carried on, such as cotton-spinning, machine-building, brass-
founding, colour-printing, tobacco-making, tanning, brewing, and
blouse-making. The Corporation carries on the gas-making,
waterworks, electric lighting, and trams. In addition to the
usual public buildings it possesses Museums of Art and Natural
History, and University College. (pp. 1, 5, 12, 40, 41, 50, 73,
74, 77, 88, 89, 100, 102, 103, 108, 110, 112, 114, 115, 118,
127–132.)

Ollerton (690), on the Maun, nine miles N.E. of Mansfield,
near the northern border of Sherwood Forest, is a convenient
centre from which to visit the Dukeries. Close by are the "Birk-
lands," so called on account of the great number of birches.
There are also two famous oaks here, the Major Oak and Robin
Hood's Larder. (pp. 32, 103.)

Oxton (455), a village five miles S.W. of Southwell, interesting as having in its proximity some of the best preserved

earthworks in the county. Those known as Oldox have multiple ramparts and seem to be of Celtic origin. West of this is a rectangular encampment, also several mounds, of which one was certainly a tumulus. Recently a boring was made near here which struck workable coal at a depth of over 2000 feet. (pp. 67, 80, 81.)

Radford (43,933). A busy manufacturing centre incorporated in the borough of Nottingham 1877.

Retford or **East Retford** (12,340). On the Idle, seven miles east of Worksop. As a borough it comes next in antiquity to Nottingham, and is said to have been incorporated by Richard I. It had the privilege of sending representatives to Parliament as early as 1315. After fifteen years had elapsed this privilege was not again used until 1571. It now belongs to the Bassetlaw division. It is head of a Poor Law Union and Petty Sessional Division. It is on the Great Northern main line and was formerly a post-town on the Great North Road. It has iron foundries, paper-mills, and india-rubber works, and is still an important market town. (pp. 60, 67, 71, 103, 112, 114.)

Scrooby (181). Now a quiet village on the winding Ryton two miles south of Bawtry. At one time there was a great palace here, the seat of the archbishops of York. Later the place became an important post-town on the Great North Road. William Brewster, who was postmaster, lived in the remains of the palace. He was the leader of the Pilgrim Fathers who, unable to gain religious liberty in England, left for Holland in 1607. From thence they sailed in the *Mayflower* in 1620 and founded a settlement in Virginia, which province was then under the Governorship of Sir Edwin Sandys, a brother of the owner of Scrooby. (pp. 46, 115, 116, 126.)

Selston (7071). A colliery town seven miles S.W. of Mansfield. It seems to have been the first place at which coal was

worked in this county. In the fifteenth century the monks of
Beauvale had a coal-mine here.

Shelford (386) is six miles E.N.E. of Nottingham and is a
good example of a village built upon a gravel patch.   In the
illustration here given the foreground is occupied by the low-
lying alluvium.   At the back the church, cemetery, and village
can be seen standing at a higher level.   (pp. 89, 122.)

Church and Village of Shelford

Sneinton or **Snenton** (23,093). Incorporated in the
borough of Nottingham 1877.   At one time there were a
number of cave dwellings here.

Southwell (3161).   On the Greet.   A historic little place
which owes its importance almost entirely to the presence of the
splendid cathedral and the consequent procession of ecclesiastics
and their trains.   It was made the seat of a bishopric in the
sixteenth century but lack of funds caused this to be abolished.

S. N.   10

In 1884 it was reestablished and made the head of the diocese of Nottinghamshire and Derbyshire. The Saracen's Head Inn is noted for its associations with Charles I. At one time Southwell was a busy market town, but in common with several other places the market has practically disappeared because of the facilities of access to such places as Newark and Nottingham, and easy delivery of goods from these towns by van as well as rail. It is head of a Poor Law Union and Petty Sessional Division. (pp. 60, 66, 84–87, 90, 98, 102, 112, 114, 115.)

Stapleford (5766), six miles west of Nottingham, has not only the Hemlock Stone, p. 79, but its cross is the oldest Christian memorial in the county, believed to date from A.D. 680 to 780. It is marked with wonderfully intricate scroll-work decoration.

Stoke or **East Stoke**, on the Trent four miles S.W. of Newark, is noteworthy as the scene of the battle between the forces of Henry VII and the impostor Lambert Simnel. (p. 74.)

Sutton-in-Ashfield (14,862), three miles S.W. of Mansfield, is primarily a colliery town. Hosiery manufacturing has advanced rapidly, and there are important stock fairs. Many persons also find employment in connection with the railway. (pp. 60, 103, 122.)

Thoresby, near Ollerton. The seat of Earl Manvers. The original mansion, in which Lady Mary Wortley Montagu was born, was destroyed by fire in 1745. The present building of Steetley stone is Elizabethan in style. (p. 99.)

Thurgarton, three miles south of Southwell, was formerly the site of an Augustinian priory. The parish church was the church of the priory founded in 1130. (pp. 88, 90, 111.)

Tuxford (1283). Formerly a post-town on the Great North Road; seven miles south of Retford, and until recently a market town of some importance in a rich agricultural district. Now only the cattle-market is held, the general market having succumbed to the facilities of railways and vans. The town is on the Great Northern main line, which is here crossed by a branch of the Great Central Railway. (pp. 49, 53, 103, 124.)

Welbeck (97). Welbeck Abbey, near Worksop, is the seat of the Duke of Portland. It is built on the site of a Praemonstratensian abbey, only fragments of which remain. Close by is the romantic gorge known as Creswell Crags. (pp. 7, 49, 75, 90, 99.)

The Park, Wollaton Hall

West Stockwith (667). A small river port situated in the extreme north of the county at the point where the Idle, the Morther drain, and the Chesterfield canal enter the Trent. Large chemical and engineering works provide many of the people with employment. (pp. 16, 60.)

Willoughby - on - the - Wolds (398), some seven miles E.N.E. of Loughborough, is the original home of the Willoughby family. Close to this village, but on the Fosseway, was the Roman station Vernometum or Verometum. A battle was fought here during the civil wars in 1648. (p. 71.)

Wollaton (541), three miles west of Nottingham, is chiefly noteworthy for Wollaton Hall, the country seat of Lord Middleton. It was built by John of Padua from 1580–88 at a cost of some £80,000 and is a magnificent example of an Elizabethan mansion. Ancaster oolite was used and was obtained in exchange for coal. (pp. 61, 62, 98, 102, 117.)

Worksop (16,112). Here was once an Augustinian priory which is now in ruins. The beautiful Norman church, however, remains, and is one of the finest in the county. There was also a famous manor house. Around these the town grew up and is still a busy market town. Malting, timber-sawing and wood-work, especially the making of Windsor chairs, are the chief industries. It is head of a Poor Law Union and Petty Sessional Division. Worksop Manor, which formerly belonged to the Dukes of Norfolk, is now the property of Sir John Robinson. (pp. 54, 60, 90, 103, 112, 114, 122.)

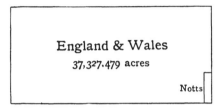

Fig. 1. The Area of Nottinghamshire, 539,756 acres,
compared with that of England and Wales

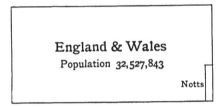

Fig. 2. The Population of Nottinghamshire (514,578)
compared with that of England and Wales in 1901

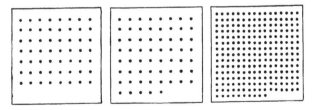

England & Wales, 558   Nottinghamshire, 610   Lancashire, 2347

Fig. 3. Comparative Density of Population per square
mile in 1901

(Each dot represents ten persons)

10—3

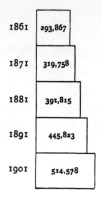

Fig. 4.   Growth of Population in Nottinghamshire
for each Census from 1861

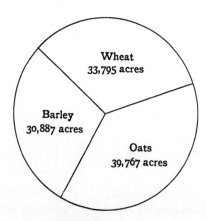

Fig. 5.   Proportionate Area of chief Cereals in
Nottinghamshire in 1907

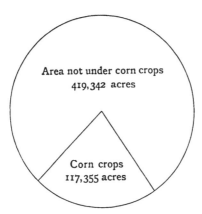

Fig. 6.   Proportionate Area under Corn Crops in
Nottinghamshire in 1907

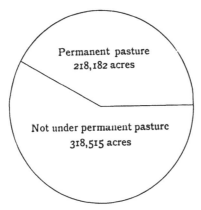

Fig. 7.   Proportion of Permanent Pasture in
Nottinghamshire in 1907

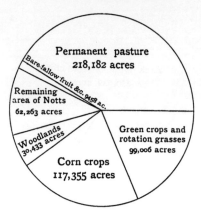

Fig. 8.   Proportion of Permanent Pasture to other
Areas in Nottinghamshire in 1907

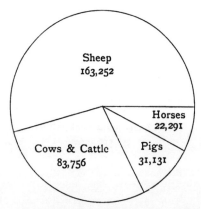

Fig. 9.   Proportionate numbers of Live Stock in
Nottinghamshire in 1907

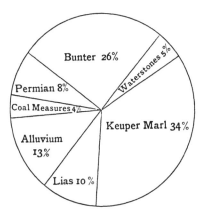

Fig. 10. Proportionate representation of the Outcrop of
Geological Formations in Nottinghamshire